# MATHS FOR ADVANCED BIOLOGY

**Alan Cadogan**
B.Sc., F.L.S., C.Biol., F.I.Biol.
Teacher, Lecturer, Chief
Examiner at A-Level and
Examiner of projects
at A/AS-Level

**Robin Sutton**
B.Sc., Ph.D.
Ecology Tutor, Field Studies
Council, Rhyd-y-creuau,
Betws-y-coed

Nelson

# *Contents*



---

Thomas Nelson and Sons Ltd
Nelson House Mayfield Road
Walton-on-Thames Surrey
KT12 5PL UK

Nelson Blackie
Wester Cleddens Road
Bishopbriggs
Glasgow G64 2NZ

Thomas Nelson (Hong Kong) Ltd
Toppan Building 10/F
22A Westlands Road
Quarry Bay Hong Kong

Thomas Nelson Australia
102 Dodds Street
South Melbourne
Victoria 3205 Australia

Nelson Canada
1120 Birchmount Road
Scarborough Ontario
M1K 5G4 Canada



Cover and text photographs courtesy of Graham Burns Photography

First published by Thomas Nelson and Sons Ltd 1994

ISBN 0-17-448214-0
NPN 9 8 7 6 5 4 3 2



Printed in China

# Introduction

In his autobiography Charles Darwin wrote

*'During the three years which I spent at Cambridge my time was wasted. I attempted mathematics but I got on very slowly. The work was repugnant to me, chiefly from my not being able to see any meaning in the early steps - this was very foolish and in after years I deeply regretted that I did not proceed far enough to at least understand something of the great leading principles of mathematics; for men thus endowed seem to have an extra sense.'*

Darwin's comment no doubt rings true to many biologists working today, but in fact it is no longer possible to study natural science without some background understanding of the skills and techniques of mathematics.

This book aims to present, in a simple and straightforward manner, some of the aspects of number work you are likely to encounter in your advanced studies of biology. Your course is almost certain to involve investigative work which may take the form of a biological problem to be solved in the laboratory or while on field-work, or a study of some aspects of human and social biology and genetics.

Scientific books and papers are generally supported by data in the form of tables and graphs. This book explains some of the principles of data handling, concentrating on the statistics that will be useful to you in your general reading of science and in your own research investigations.

Not all examination boards have the same requirements - you will need to check the regulations to see exactly what is expected of you. Some examining authorities expect a concise investigation - where you look at a problem and collect data in a few hours - others expect you to work on your own over an extended period of time. In both cases you will be expected to present the work in a neat and scientific form. Neatness doesn't mean that it *must* be word-processed, though access to a personal computer can certainly be useful. You won't receive more marks for a word-processed report, but it is a useful way for you to store, change and edit your work until you are satisfied with the words, the shape of paragraphs and the sequences of the report.

Note that in Higher Education, students are expected to use computers in data analysis. There are some programs available that are suitable for advanced level students.

Early on in the text you will find a list of investigations which may prompt you to think of a topic for yourself, perhaps related to something you have already observed. There are also a number of case studies - real data produced by students in the field. These will act as guides for your statistical treatment of data.

It is important, though, to always be aware of what statistics actually show; a famous phrase claims that scientists should use statistics 'like a drunken man uses a lamp-post; for support rather than for illumination'!

Alan Cadogan

Robin Sutton

# Chapter 1 Carrying out an investigation

*Erica cutting back ivy on mature trees*

## Case study 1

Erica, a student, was taking part in a conservation day at her local nature reserve. She was given the task of hacking down the ivy growing over the mature trees in a woodland. In the midst of this thankless task she became aware that the ivy leaves growing on the outer sunny edge of the wood seemed smaller than those growing in the inner shady areas of the wood. She put down her machete, selected four sites with decreasing amounts of light and labelled them A, B, C and D.

Erica had an idea that the size of the ivy leaves in the four sites increased in the sequence

A –> B –> C –> D.

She started her investigations by roughly measuring the width of five leaves in each site. From her results it was obvious that the leaves from site A were smaller than the leaves from site D, but she could not be sure about the whole sequence.

As she worked through the afternoon Erica realised that to continue her investigation she needed to ask herself some questions:

- what would be a suitable method of measuring the amount of light falling on the different ivy plants?
- how could she be sure that the light wouldn't change as she went from plant to plant?
- how many different areas should she record to give a good range of light?
- what would be the best indication of leaf size - thickness, length, width, area or mass?
- how many leaves should she check?
- should the leaves all come from one height?
- how could she avoid selecting the biggest leaves?
- if she expected a site to have smaller leaves would she select small examples without realising it?
- having measured a number of leaves and recorded the raw data what would be a suitable way to describe any difference found?
- if she found some difference, could she be sure it was a *real* difference or merely due to chance?
- how could she write a report to convince someone that she had made a useful ecological discovery?

Erica now had an idea as to how best to carry out her own investigative assignment and it had all come about from her initial experience and interest.

This is usually the best way to start - try to find something that really interests you. You may notice something odd, like different vegetation on each side of a hedge, and decide to look into it further. You may doubt something that you have read, for example that some varieties of yeast produce more alcohol than others, and decide to test the statement. You may have carried out a coursework experiment and decide to test a range of conditions. Ecological work can be a rich source of ideas for investigations, particularly for students on field courses.

We will return to Erica and her investigation of the ivy leaves later. First let's consider some general points on how to prepare and write up project investigations.

## THE STANDARD INVESTIGATION REPORT

A scientific report should follow a standard form as summarised below:

**Title:** You may have to present a working title before you complete the work so make sure that the title states clearly and briefly what your report is about. Remember that it is a scientific report - so a tabloid newspaper type headline is not acceptable. Avoid jargon and attempts at humour.

**Abstract:** All scientific papers open with an abstract, i.e. a brief paragraph which gives potential readers enough information to decide whether or not to spend time reading the whole paper. It should contain a couple of sentences that explain what was being investigated, the method used, a simple statement of the results and general conclusions.

COMPARISON OF REGIONS OF SCHOOL FIELD

| SPECIES | AREA A | AREA B |
|---|---|---|
| GRASS A (F. RUBRA) | 𝍸 | III |
| GRASS B (F. OVINA) | III | |
| GRASS C (?) | II | 𝍸 𝍸 I |
| DANDELION | I | III |
| PLANTAIN | | 𝍸 I |
| CHICKWEED | | II |

*Figure 1.1 A page from a field notebook*

**Introduction:** This is the *real* start of the report. It may have to set the scene if the report is dealing with a fieldwork assignment, or explain the background if it is a follow-up of some theoretical work. It may be the best place to develop and state the idea (hypothesis) being tested.

**Method:** Describe clearly what you did, giving enough detail for someone else to repeat the work using your account. Explain your techniques and how you made or used any special apparatus. Details of 'recipes' for solutions or explanations of standard techniques should be put in an *appendix*. An appendix could also be used for your raw data, you could even include the actual pages of your field notebook or lab-book (see Fig.1.1). Don't be timid about giving details of things that didn't work - use them to show how your thinking and technique evolved. Describe any pilot experiments or trial runs. Explain how you carried out control experiments and how you ensured that you had taken a suitable sample and carried out replicate tests.

**Results:** This is where you use, not the raw data, but a presentation of your results as tables, graphs or charts. It is very important to explain the results (see the advice on data presentation on page 5). If you carried out statistical tests on your data the results could be given here and the workings included as an appendix. Make sure that you explain what any statistical tests indicate and how you can use them to support or reject your hypothesis.

**Conclusions:** This is the part of the report that demonstrates the quality of your thinking. State each conclusion clearly, together with evidence and any limitations. The conclusions should be drawn directly from your work and your results and, ideally, should relate back to your introductory statement of the problem or the hypothesis you set out at the start.

**Limitations and modifications:** After carrying out an investigation you will be aware that you have been limited by available time, lack of technique or suitable apparatus, etc. State these limitations. You may also be able to suggest modifications that could be made by someone else starting to do a similar investigation. Give details and examples of any further work you would do if you continued the investigation.

**Acknowledgements:** This section gives you an opportunity to acknowledge any help received from colleagues, tutors, technicians or those you may have written to for help. It is also the place for the *bibliography*.

**Bibliography**: This is a descriptive list of books that a reader of your report might find useful to read. You may also have used these books as a reference when writing your report. The references must be given in the standard form i.e. author, date, title, (in italics or underlined), publisher. The authors should be listed in alphabetical order. Here are some examples:

Cadogan, A. and Best, G. (1992). *Environment and Ecology*, Nelson
Chalmers, N. and Parker, P. (1986). *The OU Project Guide*, Field Studies Council
Olejnik, I. and Farmer, B. (1989). *Practical Biotechnology for A-Level*, Nelson

## BEFORE YOU START YOUR INVESTIGATION

Before you carry out any investigation there are a number of questions you should ask yourself.

- have you selected a topic that interests you and about which you can be enthusiastic?
- are the chemicals, organisms and materials that you require readily available?
- do you need any advice, help or permission?
- is the investigation suitable in terms of safety, conservation and correct treatment of animals (Department for Education or Local Education Authority regulations)?
- if your investigation is microbiological do you know all of the appropriate safe sterile techniques?
- is your investigation suitable for the time available?
- can you find appropriate background reading to help you?
- is it a real investigation and not just a description?
- what variables are you going to consider?
- can you phrase a suitable hypothesis to test?
- will you collect data in a form suitable for statistical testing?
- have you studied the regulations and advice of your examination board? (Some boards allow you to use shared data; others insist on individual investigations.)

You should welcome the chance to do some investigation - it gives you the opportunity to tackle some real research, where you can present the problem, design the hypothesis, carry out some practical work and finally communicate your discoveries and conclusions to others in the form of a written report.

***Some topics for investigation that have been successful in the past are:***

- effect of straw-burning on species diversity.
- presence of a specific pollutant in a local habitat.
- growth of lichens on north- and south-facing roofs.
- distribution of catalase in growing seedlings.
- pectinase and yield of fruit juice.
- coppiced and non-coppiced woodland, a comparison of flora.
- effects of pH on barley seed growth.
- effects on bacteria of soap, toothpaste or deodorants.
- feeding preferences of carp.
- species diversity on two sides of a hedge.
- short-term memory and distracting music.
- colonisation of a garden pond.

*Box 1.1*

It is important to make sure that any graphs that are used in your work present your data in an accurate way and help the reader to get a quick and clear understanding of your results.

The data that you collect can be described as values of a *variable* (e.g. the yield of a reaction, the pH of soil, the speed of a response). For quantitative data variables may be *discrete* and the values are normally whole numbers (e.g. the number of shrews in a litter or the number of flowers in an inflorescence). *Continuous* variables occur if, for example, mass, length or time is being recorded and the values may not always be whole numbers.

In the course of your investigation you will often be looking at links between variables, for example does the colour of a flower depend on the soil pH, or does the mass of a starling chick increase with the number of days since hatching?

Some relationships between variables are fairly obvious. For example, parents sometimes record their children's height at every birthday. In this case the 'experimenter' selects the time variable and records a measurement every 52 weeks, so 'time' is said to be the *independent variable*. Since 'height' would seem to depend on age (i.e. time), the height measurements form a set of *dependent variables*. In another example, you might want to investigate how much yeast is produced under controlled conditions between –10° and 45°C. For this investigation you would need to select 'temperature' as the independent variable and 'yield' as the dependent variable.

### Plotting graphs

Have you ever been puzzled about which way round to plot a graph? The simple rules are
- the independent variable is plotted on the horizontal axis (the $x$-axis) .
- the dependent variable is plotted on the vertical axis (the $y$-axis).

### Types of graphs

The most common graphs are straightforward *line graphs*. Four examples of line graphs are shown in Fig.1.2. Each graph illustrates a different type of relationship.

If you look at Fig.1.2(a) you will see that as one variable increases, so does the other. This is an example of a *positive relationship*. In Fig.1.2(b) as one variable increases the other may increase or decrease giving a fluctuating variation, over a period of years. In Fig.1.2(c) as one variable increases the other increases to a point where it starts to level off. Finally, in Fig.1.2(d) you will see that as one variable (e.g. temperature) increases the other variable (e.g. yield) increases to a peak (the best yield or optimum value). With further increase in temperature the yield drops. This graph is typical of an enzyme reaction.

It is very important that you are able to explain in writing what is being shown in your graph, using words such as increasing, decreasing, maximum value, minimum value and levelling off.

*Hints on graphs*

- make sure the axes are the right way around.
- use a suitable scale to fill the available space. You can also use a broken axis line if needed (see Fig.1.3(c).
- label the axes and the units being used. (Note the use of the oblique (/) showing what is being measured (e.g. length of stem/cm) .
- give the graph a suitable title.
- points plotted should be shown as 'x' or '⊙'.
- use a sharp pencil to join points as smoothly as possible.
- use a rule to join the points if they are in a straight line (in some graphs a rule is used to join points if there is uncertainty about the exact shape of the graph between the points).

Box 2.1

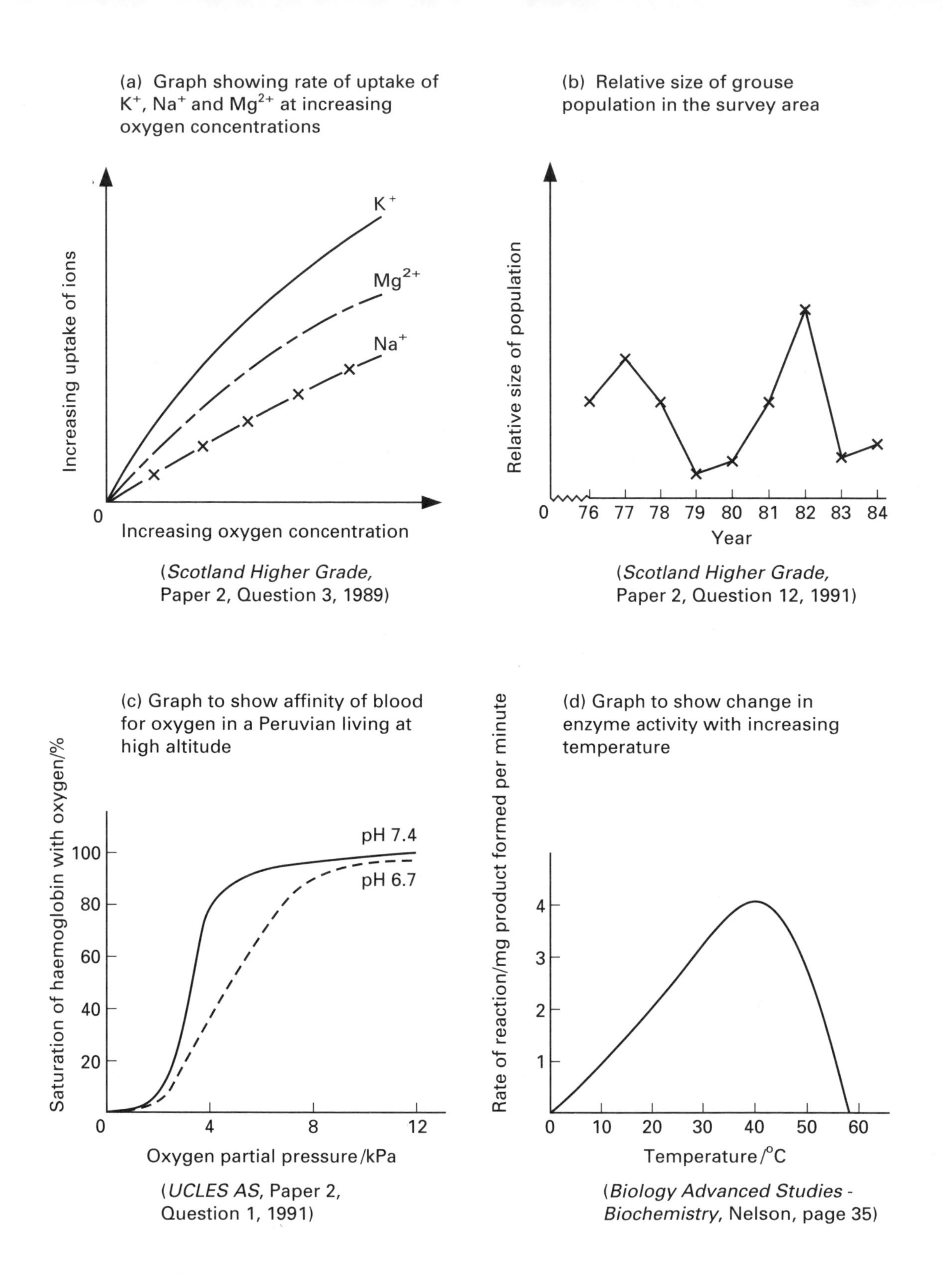


*Figure 1.2 Different types of line graph. (a) Rate of uptake of $Na^+$, $K^+$ and $Mg^{2+}$, (b) Grouse population survey, (c) Peruvian blood analysis, (d) enzyme reaction rate*

As well as line graphs other types of 'graphs' are sometimes used:

**Bar graphs**: Here there are *discrete* categories on the *x*-axis that are each shown as separate bars (see Fig.1.3(a)).

**Histograms**: Here the *x*-axis variable is *continuous* (see Fig.1.3(b)).

**Scattergrams**: Here points are plotted to show the relationship between two variables (see Fig.1.3(c)).

**Kite diagrams**: These are used to show frequency and distribution of different species. They are also called *frequency polygons* (see Fig.1.3(d)).

**Logarithmic graphs**: These are plotted on log paper and usually show rates (see page 41).

(a) Variation in the PNMR for legitimate babies born in 1979 in England and Wales, according to father's social class

(b) Number of breeding herons in the Thames drainage area (1934-1952)

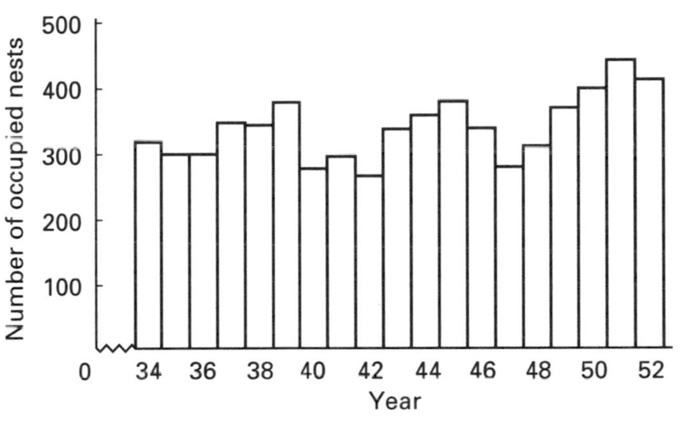


(c) Relationship between lung cancer and smoking habits for various occupational groups of men

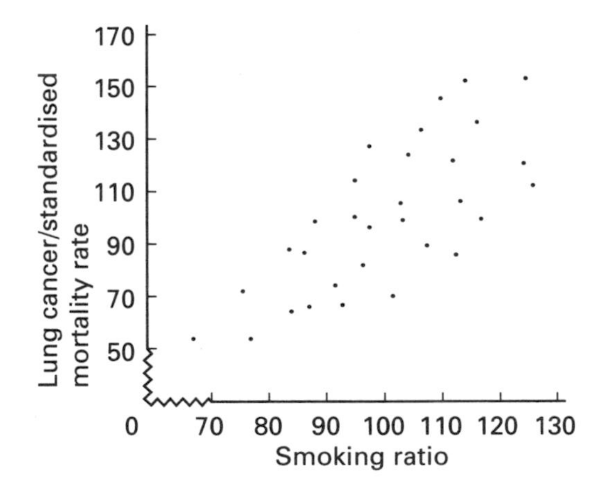


(d) Distribution of organisms in the intertidal region of a rocky shore. Distribution is represented on the y-axis. Abundance is represented on the x-axis by the width of the columns

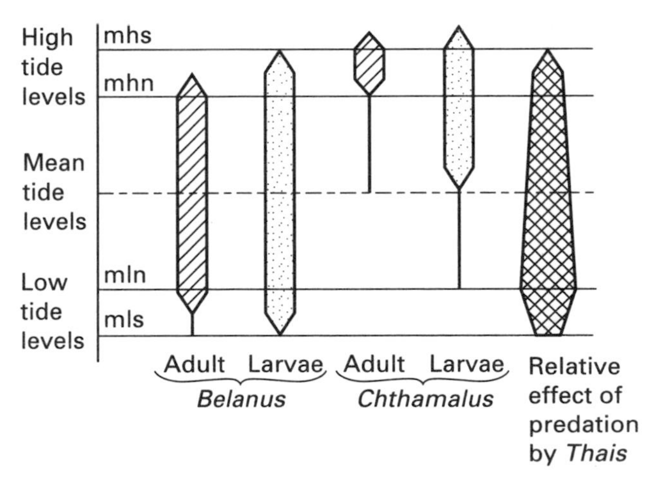


*Figure 1.3 Different types of 'graphs'*

# Chapter 2 Summarising data

## AVERAGES: MEAN, MEDIAN AND MODE

In Case Study 1 we saw how Erica measured the width of some ivy leaves. She realised that she really needed to find an *average* leaf width for each site. Instead of just five leaves she took 23 leaves from the shaded site D, measured them (in centimetres, correct to one decimal place) and arranged her results in increasing order (i.e. rank order).

Her results were as follows:

3.6, 3.8, 4.0, 4.1, 4.3, 4.4, 4.4, 4.5, 4.5, 4.5, 4.5, 4.7, 4.8, 4.8, 5.0, 5.1, 5.3, 5.4, 5.4, 5.5, 5.7, 6.4, 6.5

The word 'average' is often used when talking about samples. There are in fact three ways of measuring the average: the *mean*, the *median* and the *mode*.

**Mean**: The mean, or arithmetic mean is the sum of all the samples, divided by the number sample. It is usually written as $\overline{x}$ (see Box 2.1).

***Calculating the mean value*** $\overline{x}$

Each measurement from a sample is given the symbol $x$

The sum of all measurements is $\Sigma x$
($\Sigma$ is the Greek letter sigma)

The number of entries in the sample is $n$

The mean of the samples is $\overline{x}$ (say $x$ bar)

The equation to find the mean is as follows:

$$\text{Arithmetical mean} = \frac{\text{(Sum)}}{\text{Sample size}}$$

or $$\overline{x} = \frac{\Sigma x}{n}$$

*Box 2.1*

For Erica's results the mean is:

$$\overline{x} = \frac{\Sigma x}{n} = \frac{3.6 + 3.8 + 4.0 + 4.1 + 4.3 + 4.4, etc.}{23}$$

$$= \frac{111.2}{23} = 4.83$$

**Median**: The median of a group of results is the middle number in the list when arranged in rank order.

For Erica's results the median is the 12th number, i.e. 4.7.

Finding the median is quite straightforward if the sample size is an odd number, as in Erica's list of 23 numbers. If the sample size is even, for example 24 numbers, the median is taken to be the arithmetic mean of the middle two numbers (in this case the 12th and 13th entries) in the ranked data. In the case of both an even- and odd-numbered set there should be as many values above the median as below it.

**Mode**: Another value you might want to calculate from your results is the mode, or the measurement which occurs the greatest number of times.

For Erica's results the mode is 4.5 as it occurs four times.

Sometimes there may be several modal values in a set of data. Look at the list of numbers below:

3.6, 3.8, 4.0, 4.0, 4.1, 4.2, 4.2, 4.2, 4.3, 4.3, 4.4, 4.4, 4.4, 4.6.

There is an even number of entries in the sample ($n$=14). The mean is 4.2, the median is 4.2, but the mode has two values, 4.2 and 4.4.

Erica wanted to determine if her 23 leaf widths from site D were normally distributed. Her data ranged from 3.6 to 6.5 cm. She split this range into a number of equal size classes and counted up (tallied) the number of leaves falling into each class. From this she was able to produce a histogram (see Fig.2.1). Erica's results are 'bell-shaped' with the data spread evenly around the mean, i.e. a typical *normal distribution curve.*

| Size class/cm | Tally |
|---|---|
| 3.46 - 3.95 | II |
| 3.96 - 4.45 | 𝍸 |
| 4.46 - 4.95 | 𝍸 II |
| 4.96 - 5.45 | 𝍸 |
| 5.46 - 5.95 | II |
| 5.96 - 6.45 | I |
| 6.46 - 6.95 | I |

*Table 2.1*

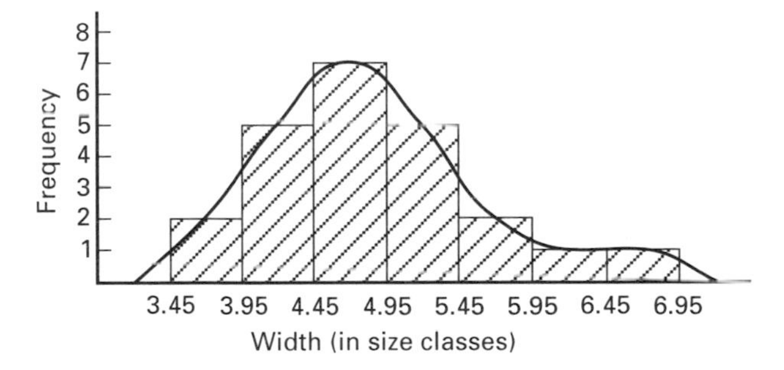


*Figure 2.1 Erica's data*

*Figure 2.2 Normal distribution curve*

Most people, if asked to summarise a set of data, would probably come up with the idea of using some sort of average, they might even be able to describe how to calculate the mean value. For data that seem to be normally distributed this is an entirely satisfactory start. If asked what other information was needed some might suggest that a measure of the 'spread' of the data would be useful. The simplest way of doing this is to record the *range*. So a basic summary would consist of the mean, and a record of the smallest and largest values.

This basic summary, however, can be misleading. The mean and range for two sets of data could be exactly the same yet the distribution of data within the range could be completely different, as shown in Fig.2.4 on page 13.

Erica decided to measure the *deviation* of each measurement from the mean. If her data are clustered around the mean she would expect to have lots of small *deviations* from this mean. If her data are more evenly spread over the range she would expect the deviations to be much bigger. Erica thought it might be useful to add all these deviations to give an overall measure of how tightly clustered the data are. Unfortunately, for every negative difference there is a positive one (remember the normal distribution is symmetrical) so the sum would be zero and therefore not a very useful summary!

A simple mathematical way to get rid of the positive and negative signs is to square the deviations. It is then possible to add up (sum) the squared deviations to give an idea of the spread of the data. If there are lots of values then the final sum will be big, even if all the measurements are very close to the mean. To avoid this confusion we divide the final sum by the number of measurements taken. Lastly, because the initial deviations were squared, we need to take the square root of the result to find the *standard deviation* (see Box 2.2).

***Calculating the standard deviation s***

***Method 1***

*Step 1* Calculate mean value $\overline{x} = \frac{\sum x}{n}$

For Erica's data from site D, $\overline{x}$ = 4.83 cm

*Step 2* Measure the deviations $x - \overline{x}$

*Step 3* Square the deviations $(x - \overline{x})^2$

*Step 4* Add the squared deviations

$\sum(x - \overline{x})^2 = 12.3327$

*Step 5* Divide by the number of samples

$$\frac{\sum(x - \overline{x})^2}{n} = \frac{12.3327}{23}$$

*Step 6* Take the square root

$$s = \sqrt{\frac{\sum(x - \overline{x})^2}{n}} = \sqrt{\frac{12.3327}{23}} = 0.7323$$

Therefore, the standard deviation $s = 0.7323$

| **Leaf** $n$ | **Width/cm** $x$ | **Width – mean** $x - \overline{x}$ | **(Width – mean)$^2$** $(x - \overline{x})^2$ |
|---|---|---|---|
| 1 | 4.5 | -0.33 | 0.1089 |
| 2 | 5.0 | 0.17 | 0.0289 |
| 3 | 4.5 | -0.33 | 0.1089 |
| 4 | 4.4 | -0.43 | 0.1849 |
| 5 | 4.5 | -0.33 | 0.1089 |
| 6 | 4.7 | -0.13 | 0.0169 |
| 7 | 4.8 | -0.03 | 0.0009 |
| 8 | 6.4 | 1.57 | 2.4649 |
| 9 | 4.4 | -0.43 | 0.1849 |
| 10 | 4.3 | -0.53 | 0.2809 |
| 11 | 5.1 | 0.27 | 0.0729 |
| 12 | 3.8 | -1.03 | 1.0609 |
| 13 | 3.6 | -1.23 | 1.5129 |
| 14 | 5.5 | 0.67 | 0.4489 |
| 15 | 5.3 | 0.47 | 0.2209 |
| 16 | 4.8 | -0.03 | 0.0009 |
| 17 | 4.5 | -0.33 | 0.1089 |
| 18 | 5.7 | 0.87 | 0.7569 |
| 19 | 4.1 | -0.73 | 0.5329 |
| 20 | 6.5 | 1.67 | 2.7889 |
| 21 | 5.4 | 0.57 | 0.3249 |
| 22 | 5.4 | 0.57 | 0.3249 |
| 23 | 4.0 | -0.83 | 0.6889 |
| Σ(Sum) | 111.2 | Σ(Sum) | 12.3327 |

*Table 2.2 Erica's results from site D.*

*Box 2.2*

Using Method 1 to calculate standard deviation is time consuming and often introduces rounding errors. Luckily there is a much quicker method which uses a different formula. The second formula is, in fact, derived from the first, and both methods give the same result.

***Calculating standard deviation s***

***Method 2***

*Step 1* Set out Erica's data as in Table 2.3 and square each value of the leaf width.

*Step 2* Sum the values of $x$ and $x^2$.

$\Sigma x = 111.2$ $\Sigma x^2 = 549.96$

*Step 3* Substitute the values of $\Sigma x$ and $\Sigma x^2$ into the equation for the standard deviation.

$$s = \sqrt{\frac{\Sigma x^2 - ((\Sigma x)^2 / n)}{n}}$$

$$s = \sqrt{\frac{549.96 - (111.2^2 / 23)}{23}}$$

$$s = \sqrt{\frac{549.96 - 537.63}{23}} = 0.7322$$

Therefore, the standard deviation $s = 0.7322$

| Leaf $n$ | Width/cm $x$ | Width² $x^2$ |
|---|---|---|
| 1 | 4.5 | 20.25 |
| 2 | 5.0 | 25.00 |
| 3 | 4.5 | 20.25 |
| 4 | 4.4 | 19.36 |
| 5 | 4.5 | 20.25 |
| 6 | 4.7 | 22.09 |
| 7 | 4.8 | 23.04 |
| 8 | 6.4 | 40.96 |
| 9 | 4.4 | 19.36 |
| 10 | 4.3 | 18.49 |
| 11 | 5.1 | 26.01 |
| 12 | 3.8 | 14.44 |
| 13 | 3.6 | 12.96 |
| 14 | 5.5 | 30.25 |
| 15 | 5.3 | 28.09 |
| 16 | 4.8 | 23.04 |
| 17 | 4.5 | 20.25 |
| 18 | 5.7 | 32.49 |
| 19 | 4.1 | 16.81 |
| 20 | 6.5 | 42.25 |
| 21 | 5.4 | 29.16 |
| 22 | 5.4 | 29.16 |
| 23 | 4.0 | 16.00 |
| Σ(Sum) | 111.2 | 549.96 |

*Table 2.3*

*Box 2.3*

In her investigation, Erica only collected a small sample of 23 leaves from site D. This small sample gave her a mean of 4.83 and a standard deviation of 0.7323. With some friends Erica went back to collect a much larger sample of 2645 leaves. She calculated the mean and standard deviation for this large sample. The mean of 4.839 was almost identical but the standard deviation of 0.7451 was considerably larger.

By taking only a small sample in her first investigation Erica had *underestimated* the standard deviation. The formulae used in Box 2.2 and 2.3 to calculate the standard deviation are designed for instances where it is possible to measure every individual within a population (every ivy leaf for instance). When only a *small sample* of the total population can be measured a slightly modified formula must be used (see Box 2.4).

*Modified formulae for calculating the standard deviation*

$$s = \sqrt{\frac{\Sigma(x-\bar{x})^2}{n-1}} \quad \text{or} \quad s = \sqrt{\frac{\Sigma x^2 - ((\Sigma x)^2 / n)}{n-1}}$$

We can prevent the underestimation of the true standard deviation by changing the divisor from $n$ to $n - 1$. For Erica's small sample of 23 leaves the standard deviation can now be more accurately recorded as 0.7487.

Box 2.4

## STANDARD DEVIATION - A WORKED EXAMPLE

### *Case Study 2*

In 1992 a group of students were studying the growth pattern of forest crops. They selected a larch plantation of known age and measured the height of 15 trees. Their results are shown in Table 2.4.

The students first checked that the data were normally distributed by splitting the range into equal size classes and tallying the number of trees in each class (see Table 2.3).

| Size Class/m | Tally |
|---|---|
| 3.95 – 4.95 | I |
| 4.95 – 5.95 | |
| 5.95 – 6.95 | II |
| 6.95 – 7.95 | II |
| 7.95 – 8.95 | III |
| 8.95 – 9.95 | II |
| 9.95 – 10.95 | I |
| 10.95 – 11.95 | II |
| 11.95 – 12.95 | I |
| 12.95 – 13.95 | I |

Table 2.3

Although the data are fairly widely spread they do seem to be symmetrically distributed around a cluster in the middle so it is fair to assume that they are *normally distributed.*

As their sample of 15 trees was small the students decided to use the modified formulae to calculate the standard deviation.

*Step 1* Square the individual values of $x$.

*Step 2* Sum the values of $x$ and $x^2$.

$\Sigma x = 138.0$ $\quad$ $\Sigma x^2 = 1345.92$

*Step 3* Substitute the values of $\Sigma x$ and $\Sigma x^2$ into the modified equation for the standard deviation.

$$s = \sqrt{\frac{\Sigma x^2 - ((\Sigma x)^2 / n)}{n-1}} = \sqrt{\frac{76.32}{14}}$$

Therefore, the standard deviation $s = 2.33$

| Tree $n$ | Height/m $x$ | (Height)$^2$ $x^2$ |
|---|---|---|
| 1 | 6.7 | 44.89 |
| 2 | 8.7 | 75.69 |
| 3 | 6.9 | 47.61 |
| 4 | 8.5 | 72.25 |
| 5 | 9.2 | 84.64 |
| 6 | 7.7 | 59.29 |
| 7 | 8.6 | 73.96 |
| 8 | 4.9 | 24.01 |
| 9 | 7.8 | 60.84 |
| 10 | 9.6 | 92.16 |
| 11 | 10.5 | 110.25 |
| 12 | 12.8 | 163.84 |
| 13 | 11.6 | 134.56 |
| 14 | 13.2 | 174.24 |
| 15 | 11.3 | 127.69 |
| Σ(Sum) | 138.0 | 1345.92 |

Table 2.4

### What does the standard deviation tell us?

The standard deviation is a measure of the variability of a set of data or, to be more precise, of its spread around the mean (see Fig.2.3). By definition about 68% of all values lie within the range of the mean plus or minus one standard deviation (i.e. $\bar{x} \pm 1s$). About 95% of all values lie within the range of the mean plus or minus two standard deviations (i.e. $\bar{x} \pm 2s$).

If the standard deviation is small then the data are clustered closely around the mean value. A large standard deviation indicates that the data are spread more widely around the mean value. If you look at Fig.2.4 you can see that the data represented in histogram (a) have a much larger standard deviation than the data shown in histogram (b).

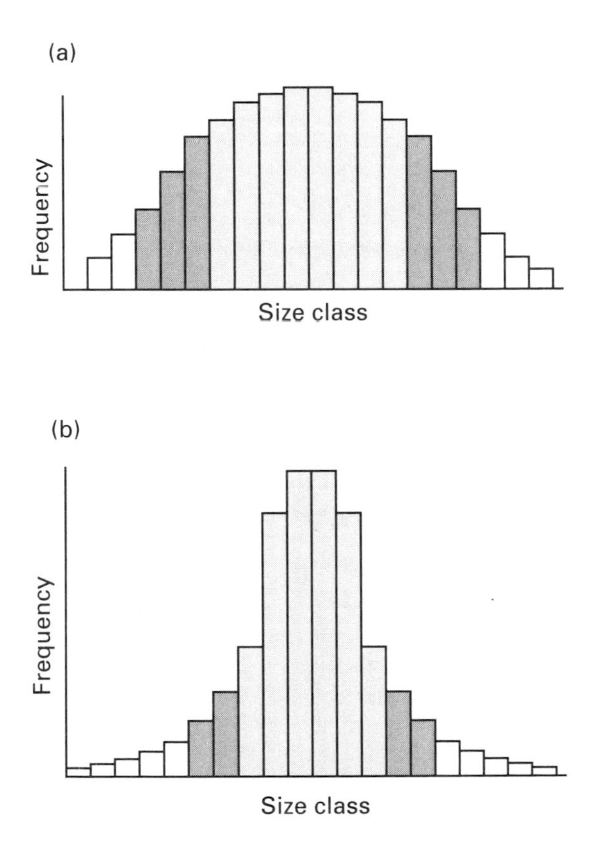


*Figure 2.4 How variable are the data? In both (a) and (b) 68% of the values lie within the range $\bar{x} \pm 1s$ and 95% of the values lie within the range $\bar{x} \pm 2s$. The values are more tightly clustered around the mean in (b) than in (a)*

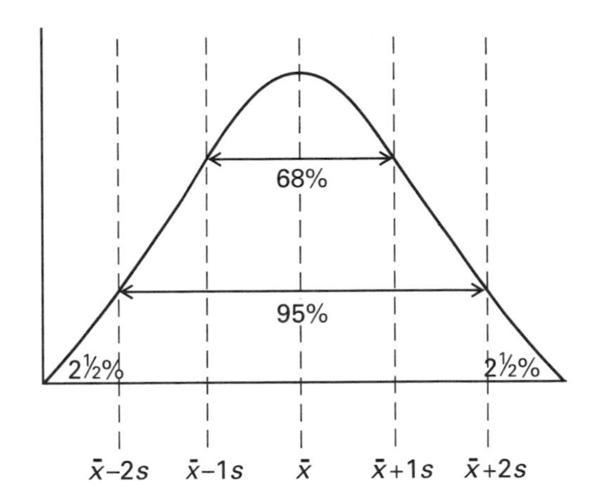


*Figure 2.3 The normal distribution curve for a large population*

### How to use standard deviations

The standard deviation is used to give an idea of the variability of a set of data and, as we have seen, it is usually presented in the form $\bar{x} \pm 1s$. The standard deviation can also be used as a graphic representation as shown in Fig.2.5. Here the mean tree heights from five different larch plantations are plotted with vertical bars representing the range $\pm 1s$.

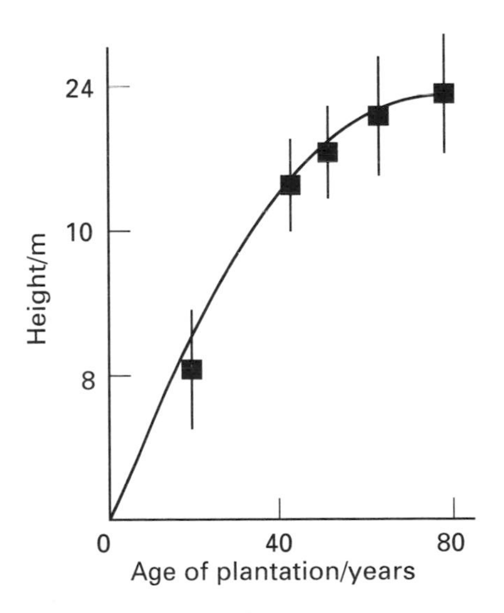


*Figure 2.5 The mean height of 15 larch trees from plantations of known age. The vertical bars represent the limit of ± 1s*

In order to explain the specific meaning of each of these terms let's look again at the *normal distribution curve* (see Fig.2.6).

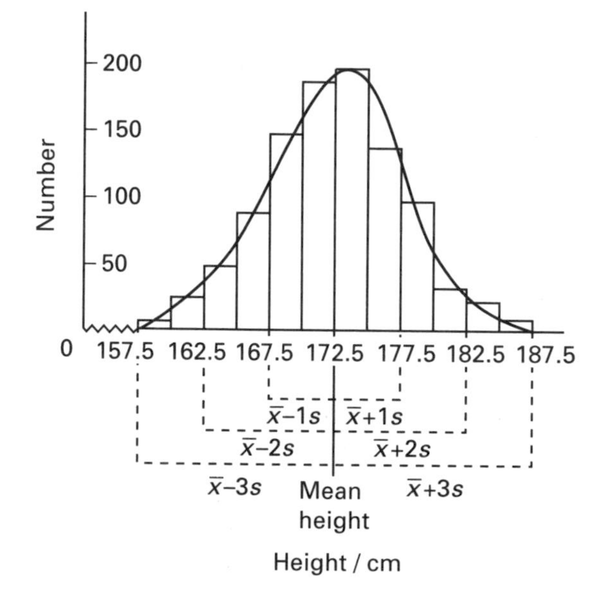


*Figure 2.6 A histogram of the heights of 1000 men (hypothetical figures) and the normal distribution curve superimposed*

### *Probability*

The divisions (i.e. $\bar{x} \pm 1s$, $\bar{x} \pm 2s$, $\bar{x} \pm 3s$) on both sides of the mean show values for one, two and three standard deviations either side of the mean. These values are called the *standard intervals* and they contain a fixed percentage of the whole population. If we work out the percentage of the population that has a height of between the mean and one standard deviation above the mean (i.e. between $\bar{x}$ and $\bar{x}+1s$), we find it is approximately 34%. Can you think why?

We can then convert this percentage into a *probability* figure, i.e. the probability of finding a person of this height in the population concerned. We give the probability of an event which is 100% certain a value of 1 and we give something which is impossible a value of 0. So, our answer of 34% is described as a probability of 0.34 (i.e. $p = 0.34$).

What do you think is the probability of finding a man who is outside the size range of between the mean and one standard deviation above the mean (i.e. outside $\bar{x}$ to $\bar{x}+1s$)?

What would you expect the probability to be of finding a person in this population outside the range of the mean plus or minus two standard deviations (i.e. outside $\bar{x} \pm 2s$)?

### *Confidence*

If you were to record the pulse rates of a group of students you might find that the mean pulse rate is 75 beats per minute, with a standard deviation of 5 beats per minute. If you can show that the distribution is normal then these results can be shown in a line chart as follows:

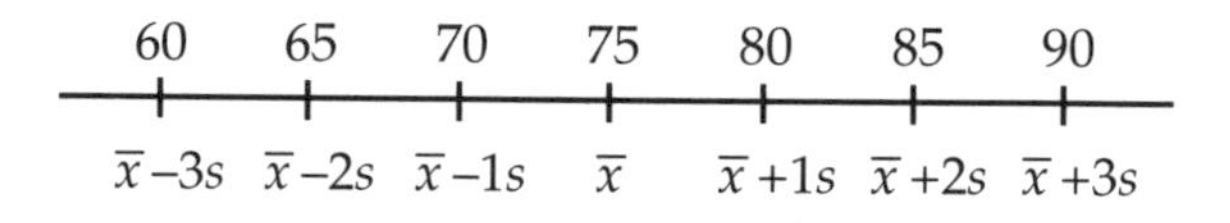


We've already seen that one standard deviation above or below the mean is approximately 34%. Therefore, $\bar{x} \pm 1s$ would contain 68% of the population. The remaining 32% would be distributed equally above and below this range. If we look back at Fig.2.3 we also see that about 95% of the population would come in the range $\bar{x} \pm 2s$. The remaining 5% would be distributed equally above and below this range.

Going back to our group of students, only 2.5% of this population could be expected to have a pulse rate lower than 65 beats per minute and 2.5% a pulse rate higher than 85 beats per minute.

So, we can be *confident* that 68% of students in the population will have a pulse rate between 70 and 80 beats per minute. If you plot on a graph a point which represents the mean of a number of variables then you could also add *confidence limits* to show the expected variability of individuals in the population. All you need to know is the value of the standard deviation for each mean (see Fig.2.5).

### *Confidence limits*

Scientists are usually cautious when they present their results! They seldom express *certainty*, but are happy to operate within a 95% confidence limit. This can be calculated by the following formula:

$$\bar{x} \pm 1.96 \frac{s}{\sqrt{n}}$$

($\frac{s}{\sqrt{n}}$ is known as the *standard error* (SE) where $s$ is the standard deviation of the sample and $n$ is the number of individuals).

In Case Study 2 we saw how students were measuring the height of 15 trees. The standard deviation was 2.33. In this case the standard error would be:

$$\frac{2.33}{\sqrt{15}}$$

Putting these values into the formula the 95% confidence limit would be:

$$\bar{x} \pm 1.96 \frac{s}{\sqrt{n}} = 9.2 \pm 1.18.$$

This value of 9.2 ± 1.18 would be shown as a vertical bar (as in Fig.2.5) but measuring 1.18 either side of the point 9.2.

The students could be 95% *confident* that the true value of the mean lay in the range 8.02 to 10.38 metres. This is called the 95% confidence limit and means that there is only a 5% chance that the real mean is *outside* this range.

### *Significance*

## Case study 3

William has started an investigation on the growth of seaweed on two shore areas, one sheltered and one exposed. He measured the lengths of 200 fronds of seaweed in each bay and found that the sample mean values from the sheltered and exposed shores differed by 6.8 cm. He wants to know if this is a real and significant difference or a mere chance occurrence.

The first thing William needs to do is to set up a *null hypothesis* (see Box 2.5), i.e. that there is no difference between the lengths of the fronds growing in the two habitats. If William can show that the probability of getting the same results *by chance* is less than 1 in 20, the null hypothesis can be rejected at the $p = 0.05$ level.

*The null hypothesis*

Is the species diversity in a stream as high below as it is above a factory outlet? Are the biceps muscles of athletes bigger than those of non-athletes? Is a 20% solution of herbicide as effective as a 10% solution? In investigations such as these the scientist, after collecting the data, must set up a *null hypothesis*. This can usually be phrased as follows:

'that there is no difference (e.g. in species diversity/biceps size/herbicidal effect) between the two populations from which the samples were taken'.

The scientist then carries out a statistical test to work out the probability of getting those values that were collected if there was no difference between the two populations (i.e. if the null hypothesis is true). If the probability is found to be low, then the null hypothesis is rejected and the scientist can say that there is a difference between the two populations. He/she can then seek reasons to explain this difference.

When is a probability described as *low*? Scientists have decided that the null hypothesis can be rejected if the probability of getting a particular set of measurements by chance is less than 1 in 20 (or less than 5%). In other words it can be rejected at the 5% significance level.

These significance levels can be described in different ways:

less than 1 in 20/5% level/$p<0.05$/or *
less than 1 in 100/1% level/$p<0.01$/or **
less than 1 in 1000/0.1% level/$p<0.001$/or ***

In this book we will be working mostly at the 5% significance level.

*Box 2.5*

# Chapter 3 Comparing two sets of data - the t-Test

One of the most common applications of statistics is to compare two sets of data, for example the heights of male and female students in a class. These heights can be represented as a frequency histogram using the same *x*-axis for both sets of data (see Fig.3.1).

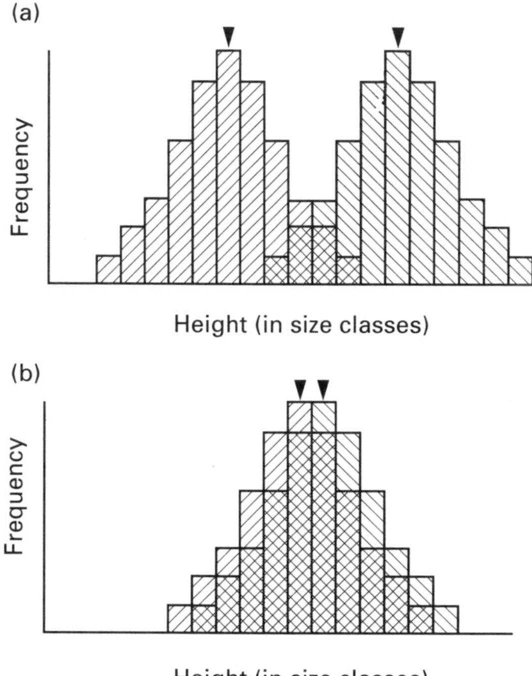


*Figure 3.1 Comparing two sets of data. The triangle indicates the size class that contains the mean value for each set of data*

If almost all the male students were taller than the female students then the two histograms would show very little overlap (Fig.3.1(a)) and we could be confident in saying that males are taller than females. As the overlap increases it becomes less certain that there is a difference. If the data looked like those shown in Fig.3.1(b) where there is almost complete overlap, then we could be confident in saying that there is no difference in the height of male and female students.

It would seem from Fig.3.1 that the difference between the mean values should be a sufficient measure of the overlap, i.e. as the means become closer the overlap increases. However, the overlap between two sets of data also depends on how closely the data are clustered around the two means. In Fig.3.2(a) the two sets of data are tightly grouped around their means and there is very little overlap. In Fig.3.2(b) the difference between the means is the same as in Fig.3.2(a) but because the two sets of data are more *variable* there is more overlap and less certainty that there is a difference between the data.

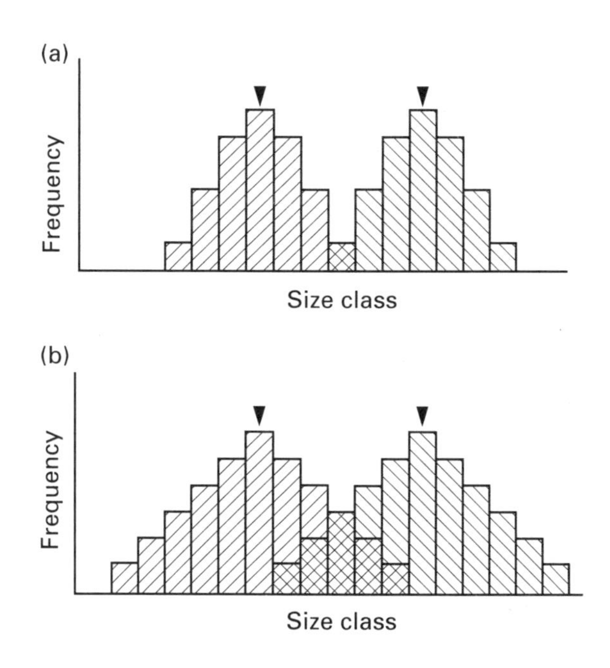


*Figure 3.2 Increasing the spread of the data also increases the overlap*

A technique is required which will measure the amount of overlap between two sets of data and say just how certain we are that there is a significant difference. The *t-Test* is one such technique and is applicable when data are normally distributed. (Otherwise we use the *Mann-Whitney U test*, see page 24.)

### *The t-Test*

$$t = \frac{|\bar{x}_1 - \bar{x}_2|}{\sqrt{\frac{s_1^{\,2}}{n_1} + \frac{s_2^{\,2}}{n_2}}}$$

The formula for the t-Test looks very complex but if you break this formula down it is easy to see it has a commonsense basis. The top part of the equation $|\bar{x}_1 - \bar{x}_2|$ compares the means of the two sets of data. This is the first factor which contributes to the overlap between two sets of data (see Fig.3.1). The bottom part of the formula includes measurements of the variability of the data ($s_1^{\,2}$ and $s_2^{\,2}$). This is the other factor which contributes to the amount of overlap (see Fig.3.2).

Let's go back to Case Study 1 and the ivy leaf investigation.

Erica measured the width of 23 leaves from the shady site D and the sunny site A. Her results are shown in Table 3.2

She had an idea that the leaves from the shady and sunny sites would differ in size, and decided to test her idea using the t-Test.

As the t-Test uses the mean and variance Erica first checked that her data were normally distributed (see Table 3.1).

| Size class/cm | Shaded site | Sunny site |
|---|---|---|
| 2.45 – 2.95 | | II |
| 2.95 – 3.45 | | 𝍸 I |
| 3.45 – 3.95 | II | 𝍸 IIII |
| 3.95 – 4.45 | 𝍸 | IIII |
| 4.45 – 4.95 | 𝍸 II | II |
| 4.95 – 5.45 | 𝍸 | |
| 5.45 – 5.95 | II | |
| 5.95 – 6.45 | I | |
| 6.45 – 6.95 | I | |

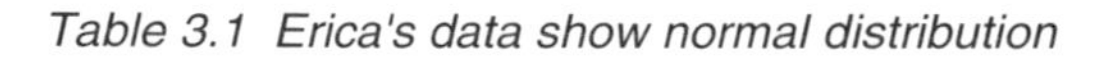
*Table 3.1 Erica's data show normal distribution*

> ***A note on notation***
>
> • $\bar{x}$ is the mean value. When comparing two sets of data, $\bar{x}_1$ and $\bar{x}_2$ are their two respective mean values.
> The vertical lines | | indicate that the positive difference between the two means should be taken, irrespective of which is bigger.
> • $s$ is the symbol for the standard deviation. The square of this value is called the *variance* $s_1^{\,2}$ and $s_2^{\,2}$ are the variances for the two sets of data and are measures of the spread of data around each mean value.
> • $n$ is statistical shorthand for the number of measurements collected. There are $n_1$ entries in the first set of data and $n_2$ entries in the second.

*Box 3.1*

| Leaf | Shaded site | | Sunny site | |
|---|---|---|---|---|
| $n$ | $x_1$ | $x_1^{\,2}$ | $x_2$ | $x_2^{\,2}$ |
| 1 | 4.5 | 20.25 | 4.5 | 20.25 |
| 2 | 5.0 | 25.00 | 3.1 | 9.61 |
| 3 | 4.5 | 20.25 | 4.1 | 16.81 |
| 4 | 4.4 | 19.36 | 3.1 | 9.61 |
| 5 | 4.5 | 20.25 | 3.1 | 9.61 |
| 6 | 4.7 | 22.09 | 3.3 | 10.89 |
| 7 | 4.8 | 23.04 | 3.5 | 12.25 |
| 8 | 6.4 | 40.96 | 3.5 | 12.25 |
| 9 | 4.4 | 19.36 | 3.8 | 14.44 |
| 10 | 4.3 | 18.49 | 3.0 | 9.00 |
| 11 | 5.1 | 26.01 | 3.6 | 12.96 |
| 12 | 3.8 | 14.44 | 3.8 | 14.44 |
| 13 | 3.6 | 12.96 | 4.5 | 20.25 |
| 14 | 5.5 | 30.25 | 2.8 | 7.84 |
| 15 | 5.3 | 28.09 | 4.2 | 17.64 |
| 16 | 4.8 | 23.04 | 3.9 | 15.21 |
| 17 | 4.5 | 20.25 | 3.4 | 11.56 |
| 18 | 5.7 | 32.49 | 3.6 | 12.96 |
| 19 | 4.1 | 16.81 | 3.6 | 12.96 |
| 20 | 6.5 | 42.25 | 3.7 | 13.69 |
| 21 | 5.4 | 29.16 | 4.1 | 16.81 |
| 22 | 5.4 | 29.16 | 4.4 | 19.36 |
| 23 | 4.0 | 16.00 | 2.8 | 7.84 |
| Σ(Sum) | 111.2 | 549.96 | 83.4 | 308.24 |

*Table 3.2 Erica's leaf width results*

### *Calculating t*

The basic steps to calculate $t$ are the same as those to calculate the standard deviation.

*Step 1* Square the individual values of $x_1$ and $x_2$.

*Step 2* Calculate the sums of the $x_1$ and $x_2$ values and of $x_1^2$ and $x_2^2$.

$$\Sigma x_1 = 111.2 \quad \Sigma x_1^2 = 549.96$$

$$\Sigma x_2 = 83.4 \quad \Sigma x_2^2 = 308.24$$

*Step 3* Calculate the means $\bar{x}_1$ and $\bar{x}_2$.

$$\bar{x}_1 = \frac{\Sigma x_1}{23} = \frac{111.2}{23} = 4.83$$

$$\bar{x}_2 = \frac{\Sigma x_2}{23} = \frac{83.4}{23} = 3.63$$

*Step 4* Calculate the variances $s_1^2$ and $s_2^2$.

$$s_1^2 = \frac{\Sigma x_1^2 - ((\Sigma x_1)^2 / n_1)}{n_1 - 1}$$

$$= \frac{549.96 - (111.2^2 / 23)}{23 - 1} = 0.56$$

$$s_2^2 = \frac{\Sigma x_2^2 - ((\Sigma x_2)^2 / n_2)}{n_2 - 1}$$

$$= \frac{308.24 - (83.4^2 / 23)}{23 - 1} = 0.26$$

*Step 5*: Substitute the mean and variances from steps 3 and 4 into the formula for $t$:

$$t = \frac{|\bar{x}_1 - \bar{x}_2|}{\sqrt{\frac{s_1^2}{n_1} + \frac{s_2^2}{n_2}}} = \frac{4.83 - 3.63}{\sqrt{\frac{0.56}{23} + \frac{0.26}{23}}}$$

$$t = 6.32$$

Box 3.2

#### *What does the t-Test tell us?*

$t$ provides a way of measuring the overlap between two sets of data. If two sets of data have widely separated means and small variances (i.e. tightly clustered data), they will have little overlap and a big value of $t$ (Fig.3.3a). If the two sets of data have means that are are close together and large variances (i.e. widely spread data), they will have a lot of overlap and a small value of $t$ (Fig.3.3b).

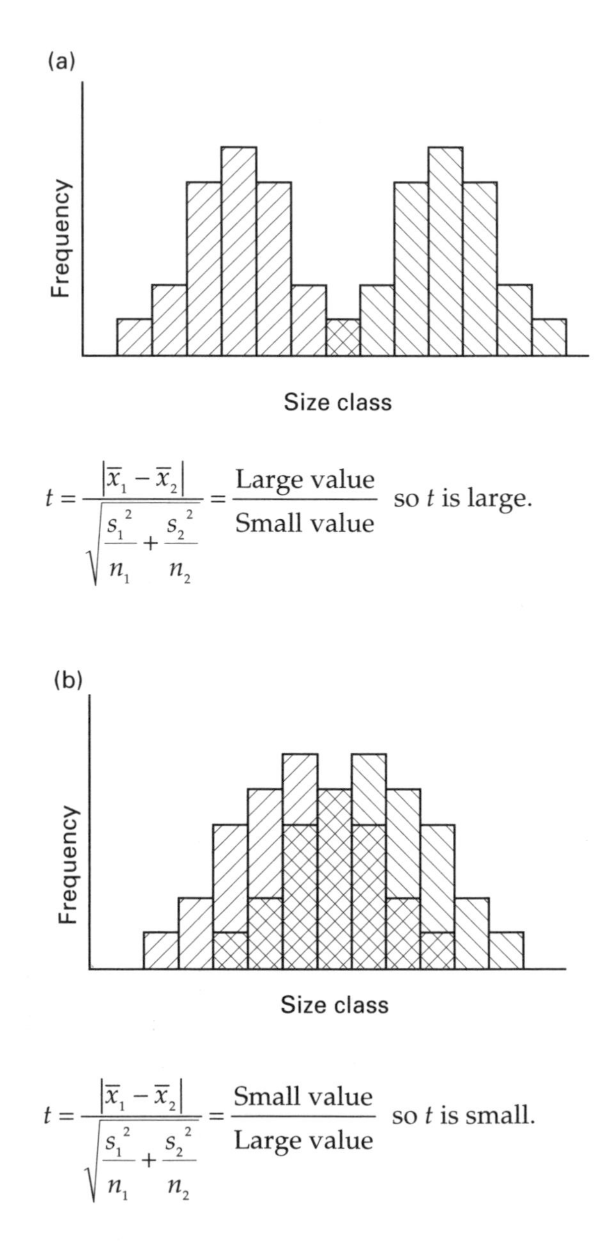


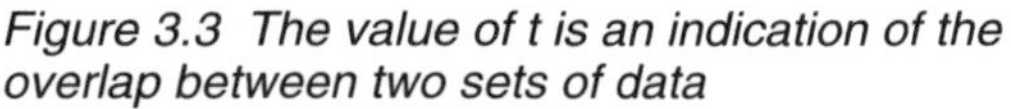

*Figure 3.3 The value of t is an indication of the overlap between two sets of data*

We have seen how a large value of *t* indicates little overlap and almost certainly a difference between two sets of data. In contrast, a small value of *t* indicates a lot of overlap and probably no difference. By picking a cut-off point on this scale it should be possible to get rid of terms like 'certainly' and 'probably' and state whether or not there is a significant difference.

This cut-off point can be found by consulting a *table of critical values* (see Table 3.3). The value that should be used from the table depends on the number of *degrees of freedom*. This is simply a way of taking into account the reliability of the data by relating the critical value to the number of measurements collected (see Box 3.3).

| Degrees of freedom | Significance levels p = 0.05 | p = 0.01 |
|---|---|---|
| 15 | 2.13 | 2.94 |
| 16 | 2.12 | 2.92 |
| 17 | 2.11 | 2.90 |
| 18 | 2.10 | 2.88 |
| 19 | 2.09 | 2.86 |
| 20 | 2.09 | 2.85 |
| 21 | 2.08 | 2.83 |
| 22 | 2.07 | 2.82 |
| 23 | 2.07 | 2.81 |
| 24 | 2.06 | 2.80 |
| 25 | 2.06 | 2.80 |
| 30 | 2.04 | 2.75 |
| 40 | 2.02 | 2.70 |
| 60 | 2.00 | 2.66 |
| ∞ | 1.96 | 2.58 |

*Table 3.3 Critical values for the t-Test. If t is greater than or equal to critical value then there is a significant difference between the two sets of data*

### *Where do critical values come from?*

It might help to think of a statistician sitting at a big computer asking it to generate random numbers (e.g. 23 width values for leaves in the shade and 23 width values for leaves in the sun). From this data he calculated a value of *t*, then asked for a second set of data and repeated this exercise. He did this calculation hundreds of times for different numbers and found that he got a value of 2.02 or more only one time in twenty (i.e. 5% or p = 0.05). This value of 2.02 represents the critical value at p = 0.05 for 40 degrees of freedom, i.e. the nearest number to the value given by the formula $n_1 + n_2 - 2$, where 23 + 23 – 2 = 44.

In her investigation Erica calculated *t* as 6.32. As this value is greater than the critical value of 2.02 (at a level of p = 0.05), Erica concluded that there was a significant difference between her two sets of data.

In fact, Erica's value of *t* was greater than the critical value of 2.70 at a p = 0.01 level, so there was less than a one in one hundred chance of the difference being due to chance.

***Using the t-Test***

- it is used to test the difference between two sets of data.
- it should only be used on data that are normally distributed.
- it should only be used ideally with large samples (> thirty measurements per set of data) and the value of *t* should be compared with the critical value at ∞ degrees of freedom.
- for smaller samples (less than thirty measurements per set of data) the value of *t* is only approximate and the degrees of freedom should be found using the formula $n_1 + n_2 - 2$
- if *t* is *greater than or equal to* the critical value, then it is possible to reject the null hypothesis and accept that there *is a significant difference* between the two sets of data.
- significant results are often shown as probability levels or denoted by asterisks, e.g. $p < 0.05$ or * $p < 0.01$ or **
- non-significant results are often shown as probability levels or denoted by initials e.g. $p > 0.05$ or n.s.

*Box 3.3*

### Case study 4

Ian was comparing the ground flora of two woods, one deciduous and one coniferous. He noticed that in the deciduous wood there seemed to be more light reaching the ground than there was in the coniferous wood. Realising the ecological importance of light Ian decided to measure the light intensity at 15 randomly selected spots within each wood. His light meter showed the values as arbitrary units (see Table 3.4).

He wanted to use the t-Test on the data, so first he checked that the data were normally distributed. He split the range of light intensities into equal size classes and tallied the number of readings in each class. His data appeared to have a normal distribution. (Try to prove this yourself!)

Ian then formulated the null hypothesis that there is no significant difference between the light intensities in the two woods.

| Site | Deciduous | | Coniferous | |
|---|---|---|---|---|
| $n$ | $x_1$ | $x_1^2$ | $x_2$ | $x_2^2$ |
| 1 | 10.5 | 110.25 | 9.1 | 82.81 |
| 2 | 9.6 | 92.16 | 10.3 | 106.09 |
| 3 | 10.1 | 102.01 | 10.8 | 116.64 |
| 4 | 11.6 | 134.56 | 10.3 | 106.09 |
| 5 | 11.6 | 134.56 | 9.6 | 92.16 |
| 6 | 11.3 | 127.69 | 11.1 | 123.21 |
| 7 | 10.6 | 112.36 | 9.3 | 86.49 |
| 8 | 10.4 | 108.16 | 10.5 | 110.25 |
| 9 | 12.4 | 153.76 | 10.4 | 108.16 |
| 10 | 11.3 | 127.69 | 9.7 | 94.09 |
| 11 | 10.7 | 114.49 | 10.2 | 104.04 |
| 12 | 10.5 | 110.25 | 10.2 | 104.04 |
| 13 | 11.5 | 132.25 | 9.7 | 94.09 |
| 14 | 11.1 | 123.21 | 10.9 | 118.81 |
| 15 | 11.0 | 121.00 | 9.9 | 98.01 |
| Σ(Sum) | 164.2 | 1804.40 | 152.0 | 1544.98 |

*Table 3.4 Ian's light intensity readings*

*Step 1* Square the individual of readings $x_1$ and $x_2$.

*Step 2* Calculate the sum of the $x_1$ and $x_2$ values and of the $x_1^2$ and $x_2^2$ values.

$$\Sigma x_1 = 164.2 \quad \Sigma x_1^2 = 1804.40$$
$$\Sigma x_2 = 152.0 \quad \Sigma x_2^2 = 1544.98$$

*Step 3* Calculate the means $\bar{x}_1$ and $\bar{x}_2$.

$$\bar{x}_1 = \frac{164.2}{15} = 10.95 \qquad \bar{x}_2 = \frac{152.0}{15} = 10.13$$

*Step 4* Calculate the variances $s_1^2$ and $s_2^2$.

$$s_1^2 = \frac{\Sigma x_1^2 - ((\Sigma x_1)^2 / n_1)}{n_1 - 1} \qquad s_1^2 = 0.50$$

$$s_2^2 = \frac{\Sigma x_2^2 - ((\Sigma x_2)^2 / n_2)}{n_2 - 1} \qquad s_2^2 = 0.34$$

*Step 5* Substitute the means and variances from steps 3 and 4 into the formula for $t$.

$$t = \frac{|\bar{x}_1 - \bar{x}_2|}{\sqrt{\frac{s_1^2}{n_1} + \frac{s_2^2}{n_2}}} = \frac{|10.95 - 10.13|}{\sqrt{\frac{0.50}{15} + \frac{0.34}{15}}}$$

$$t = 3.42$$

*Step 6* Calculate the degrees of freedom using the formula $n_1 + n_2 - 2$. (i.e. $15 + 15 - 2 = 28$). The nearest degrees of freedom therefore is 30.

Ian's value of $t$ (i.e. 3.42) is greater than the critical value (i.e. 2.04) at the $p = 0.05$ level. Therefore there is a significant difference between Ian's two sets of data. In fact, Ian's value of $t$ was greater than the critical value of 2.75 at the $p = 0.01$ level, so there was a one in one hundred chance of the difference being due to chance.

The null hypothesis is rejected ($p < 0.05$).

# Chapter 4 Dealing with non-normal data

In Chapter 2 we looked at how to summarise data that had a normal distribution. However, a lot of data, especially the small sets of data you are likely to collect in your project work, will not have this type of distribution. Two examples of non-normal data are given in Box 4.1. In both examples the data are clumped towards one end of the range of measurements to give *skewed distributions*. If we use the mean to summarise data in cases like these then the extreme values tend to have a large effect.

Look carefully at the data in Box 4.1. The percentage cover of heather shows a skewed distribution. If you look at the mean value of 74.2 you will see that there are eleven values above this mean and five below it; it has been affected by the three low extreme values 9, 18 and 26. Similarly, with the stonefly numbers the mean, 22.3, as a descriptive measure, is affected by the two extreme counts 75 and 81.

For each of these skewed distributions the *median* gives a more acceptable measure of the *average* of the data. (Remember that to find the median you arrange the data in increasing order and find the middle point.)

***Examples of non-normal data***

(a) The percentage cover of heather in 16 randomly-placed quadrats.

| Quadrat number | % |
|---|---|
| 1 | 98 |
| 2 | 18 |
| 3 | 78 |
| 4 | 71 |
| 5 | 82 |
| 6 | 73 |
| 7 | 26 |
| 8 | 9 |
| 9 | 98 |
| 10 | 83 |
| 11 | 82 |
| 12 | 96 |
| 13 | 91 |
| 14 | 94 |
| 15 | 95 |
| 16 | 93 |

| Class (no.) | Tally |
|---|---|
| 0–10 | \| |
| 11–20 | \| |
| 21–30 | \| |
| 31–40 | |
| 41–50 | |
| 51–60 | |
| 61–70 | |
| 71–80 | \|\|\| |
| 81–90 | \|\|\| |
| 91–100 | 卌 \|\| |

*Table 4.1(a)*

*Averages:*
Mean = 74.2
Median = 82.5

(b) The number of stoneflies counted per 0.25 m² quadrat.

| Quadrat number | Number |
|---|---|
| 1 | 81 |
| 2 | 30 |
| 3 | 75 |
| 4 | 6 |
| 5 | 15 |
| 6 | 18 |
| 7 | 9 |
| 8 | 12 |
| 9 | 9 |
| 10 | 19 |
| 11 | 2 |
| 12 | 10 |
| 13 | 14 |
| 14 | 12 |

| Class (no.) | Tally |
|---|---|
| 0–8 | \|\| |
| 9–17 | 卌 \|\| |
| 18–26 | \|\| |
| 27–35 | \| |
| ↓–↓ | |
| 71–80 | \| |
| 81–89 | \| |

*Table 4.1(b)*

*Averages:*
Mean = 22.3
Median = 13.0

*Box 4.1*

### Case study 5

Sanjit carried out an investigation to compare the density of earthworms in two fields. By pouring a weak solution of detergent onto 0.5 m x 0.5 m quadrats positioned randomly within each field he was able to count the number of worms that came to the surface, and so calculate their density. (Sanjit used a weak solution of detergent as it is less damaging than some other reagents). He also measured the moisture and organic content of a soil sample from each quadrat (before putting on the detergent!). This all took a long time so he was only able to sample seven quadrats per field and accidentaly lost one of the soil samples. Sanjit's results are shown in Table 4.2.

Sanjit decided that as he only had a small number of samples he could not assume his data had a normal distribution so he chose the median as a summary. He arranged his data in increasing order and found the median for each set of data (see Table 4.3).

Having summarised his data Sanjit compared the two fields further. He found that there were some clear differences between the median values for each of the variables, with the number of worms giving the greatest difference. By arranging the values on a line graph he found that for each of the variables there was some overlap between the data for the two fields (see Fig.4.1).

| | **Earthworms per 0.25 m²** | | **Moisture content/%** | | **Organic content/%** | |
|---|---|---|---|---|---|---|
| **Quadrat** | **Field 1** | **Field 2** | **Field 1** | **Field 2** | **Field 1** | **Field 2** |
| 1 | 0 | 12 | 41 | 36 | 31 | 30 |
| 2 | 3 | 5 | 47 | 36 | 35 | 39 |
| 3 | 6 | 19 | 43 | 43 | 43 | 40 |
| 4 | 4 | 18 | 40 | 23 | 30 | 20 |
| 5 | 1 | 12 | 45 | 32 | 39 | 33 |
| 6 | 3 | 14 | 32 | 34 | 28 | 27 |
| 7 | 0 | 20 | 37 | – | 37 | – |

*Table 4.2 Sanjit's data. The lost data are shown as a dash (–)*

| | **Field** | **Data arranged in increasing order** | **Median** |
|---|---|---|---|
| Worms | 1 | 0 0 1 **3** 3 4 6 | 3 |
| | 2 | 5 12 12 **14** 18 19 20 | 14 |
| Moist. | 1 | 32 37 40 **41** 43 45 47 | 41 |
| | 2 | 23 32 **34** **36** 36 43 | 35 |
| Organ. | 1 | 28 30 31 **35** 37 39 43 | 35 |
| | 2 | 20 27 **30** **33** 39 40 | 31.5 |

*Table 4.3 Data from Table 4.2 arranged in increasing order. The figures in bold represent the 'middle' or median values. If there are two 'middle' values, the median is taken as the value half way between them*

(a) Number of worms

| | | | | | | | | | | | | | | |
|---|---|---|---|---|---|---|---|---|---|---|---|---|---|---|
| Field 1 | 0 | 0 | 1 | 3 | 3 | 4 | | 6 | | | | | | |
| Field 2 | | | | | | | 5 | | 12 | 12 | 14 | 18 | 19 | 20 |

(b) Moisture content

| | | | | | | | | | | | |
|---|---|---|---|---|---|---|---|---|---|---|---|
| Field 1 | | 32 | | | | 37 | 40 | 41 | 43 | 45 | 47 |
| Field 2 | 23 | 32 | 34 | 36 | 36 | | | | 43 | | |

(c) Organic content

| | | | | | | | | | | | |
|---|---|---|---|---|---|---|---|---|---|---|---|
| Field 1 | | | 28 | 30 | 31 | | 35 | 37 | 39 | | 43 |
| Field 2 | 20 | 27 | | 30 | | 33 | | | 39 | 40 | |

*Figure 4.1 Line graph showing the overlap between the data for each field*

The data on worms showed the least overlap, indicating that there was a difference in worm numbers between the two fields. There was more overlap with the moisture content, and it seemed less likely that there was a difference. There was so much overlap with the organic content, it seemed that any difference in data between the two fields was simply due to random variations between the samples.

To compare the two fields properly Sanjit needed to measure the amount of overlap between the data for each variable. To do this he used a statistical test called the *Mann-Whitney U Test*. This is similar to the t-Test but is suitable for small sets of data, or data which are not normally distributed. It makes no assumptions about the distribution of data and is therefore called a *distribution free* or *non-parametric test.*

The Mann-Whitney U Test is based on the fact that you can rank data in order of size, with the smallest value being given the lowest rank and the largest value the highest rank. Identical values are always given the same rank, i.e. the average of those ranks that are available. Some examples of how to rank data are given in Box 4.2.

***How to rank data***

(a) All values are different.

| Value | 32 | 37 | 40 | 41 | 43 | 45 | 47 |
|---|---|---|---|---|---|---|---|
| **Rank** | **1** | **2** | **3** | **4** | **5** | **6** | **7** |

Value are arranged in increasing order and ranked from 1 upwards.

(b) Some values are tied.

| Value | 0 | 0 | 1 | 3 | 3 | 3 | 3 |
|---|---|---|---|---|---|---|---|
| **Rank** | **1.5** | **1.5** | **3** | **5.5** | **5.5** | **5.5** | **5.5** |

For the two zeros the available ranks are 1 and 2. The average of these ranks is 1.5 so both zeros get a rank of 1.5. As ranks 1 and 2 have been used up the next available rank is 3, which is given to the next value of 1. The next four numbers are all 3s and the available ranks are 4, 5, 6 and 7. The average of these ranks is 5.5 so each of the 3s gets a rank of 5.5. Ranks are put in bold here for emphasis.

*Box 4.2*

(a) *Number of worms*

| | | | | | | | | | | | | | | | |
|---|---|---|---|---|---|---|---|---|---|---|---|---|---|---|---|
| **Rank ($R_1$)** | **1.5** | **1.5** | **3** | **4.5** | **4.5** | **6** | | **8** | | | | | | | **$\Sigma R_1 = 29$** |
| Field 1 | 0 | 0 | 1 | 3 | 3 | 4 | | 6 | | | | | | | |
| Field 2 | | | | | | | 5 | | 12 | 12 | 14 | 18 | 19 | 20 | |
| **Rank ($R_2$)** | | | | | | | **7** | | **9.5** | **9.5** | **11** | **12** | **13** | **14** | **$\Sigma R_2 = 76$** |

*Fig.4.2 Ranking the number of worms in Sanjit's two fields*

*Step 1* Arrange the data on the number of worms in the two fields in increasing order. Treating all the values from both fields as if they were one set of data, rank the values. The easiest way to do this is to plot the data as a simple line graph (see Figure 4.2).

*Step 2* Sum the ranks for each field $R_1$ and $R_2$.

$$\Sigma R_1 = 29 \qquad \Sigma R_2 = 76$$

*Step 3* Calculate $U_1$ and $U_2$ using the following formulae (where $n_1$ and $n_2$ are the number of samples from each field).

Field 1 $U_1 = n_1 \times n_2 + \frac{1}{2} n_2(n_2 + 1) - \Sigma R_2$

$$U_1 = 7 \times 7 + \frac{1}{2} 7(7+1) - 76$$

$$U_1 = 49 + 28 - 76$$

$$U_1 = 1$$

Field 2 $U_2 = n_1 \times n_2 + \frac{1}{2} n_1(n_1 + 1) - \Sigma R_1$

$$U_2 = 7 \times 7 + \frac{1}{2} 7(7+1) - 29$$

$$U_2 = 49 + 28 - 29$$

$$U_2 = 48$$

*Step 4* Compare the smallest $U$ value from the two fields with the critical value for the appropriate values of $n_1$ and $n_2$ (see Table 4.5). If the smallest $U$ value is *less than or equal to* the critical value then there is a significant difference between the two sets of data.

For $n_1 = 7$ and $n_2 = 7$ the critical value = 8

Sanjit's smallest U value (i.e. $U_1 = 1$) is less than the critical value (i.e. 8), therefore there is a significant difference between the number of worms found in the two fields (at the $p = 0.05$ level).

The null hypothesis is rejected ($p < 0.05$).

(b) *Moisture content*

Sanjit then ranked the values for the moisture content of the soil in the two fields. Putting the rank totals into the formula for the *Mann-Whitney U test* he found that:

$$U_1 = 34 \quad U_2 = 8$$

For $n_1 = 7$ and $n_2 = 6$ the critical value = 6

Sanjit's smallest U value (i.e. $U_2 = 8$) is greater than the critical value (i.e. 6), therefore there is no significant difference between the moisture content of the two fields (at the p = 0.05 level).

The null hypothesis is accepted ($p > 0.05$).

(c) *Organic content*

Sanjit finally ranked the values for the organic content of the soil in the two fields. Putting the rank totals into the formula for the *Mann-Whitney U test* he found that:

$$U_1 = 26 \quad U_2 = 16$$

For $n_1 = 7$ and $n_2 = 6$ the critical value = 6.

Sanjit's smallest $U$ value (i.e. $U_2 = 16$) is greater than the critical value (i.e. 8), therefore there is no significant difference between the moisture content of the two fields (at the p = 0.05 level).

The null hypothesis is accepted ($p > 0.05$).

| | |
|---|---|
| (a) Number of worms | Smallest $U = 1$ |
| (b) Moisture content | Smallest $U = 8$ |
| (c) Organic content | Smallest $U = 16$ |

*Box 4.3*

| Values of $n_1$ \ Values of $n_2$ | 1 | 2 | 3 | 4 | 5 | 6 | 7 | 8 | 9 | 10 | 11 | 12 | 13 | 14 | 15 | 16 | 17 | 18 | 19 | 20 |
|---|---|---|---|---|---|---|---|---|---|---|---|---|---|---|---|---|---|---|---|---|
| 1 | | | | | | | | | | | | | | | | | | | | |
| 2 | | | | | | | | 0 | 0 | 0 | 0 | 1 | 1 | 1 | 1 | 1 | 2 | 2 | 2 | 2 |
| 3 | | | | | 0 | 1 | 1 | 2 | 2 | 3 | 3 | 4 | 4 | 5 | 5 | 6 | 6 | 7 | 7 | 8 |
| 4 | | | | 0 | 1 | 2 | 3 | 4 | 4 | 5 | 6 | 7 | 8 | 9 | 10 | 11 | 11 | 12 | 13 | 13 |
| 5 | | | 0 | 1 | 2 | 3 | 5 | 6 | 7 | 8 | 9 | 11 | 12 | 13 | 14 | 15 | 17 | 18 | 19 | 20 |
| 6 | | | 1 | 2 | 3 | 5 | 6 | 8 | 10 | 11 | 13 | 14 | 16 | 17 | 19 | 21 | 22 | 24 | 25 | 27 |
| 7 | | | 1 | 3 | 5 | 6 | 8 | 10 | 12 | 14 | 16 | 18 | 20 | 22 | 24 | 26 | 28 | 30 | 32 | 34 |
| 8 | | 0 | 2 | 4 | 6 | 8 | 10 | 13 | 15 | 17 | 19 | 22 | 24 | 26 | 29 | 31 | 34 | 36 | 38 | 41 |
| 9 | | 0 | 2 | 4 | 7 | 10 | 12 | 15 | 17 | 20 | 23 | 26 | 28 | 31 | 34 | 37 | 39 | 42 | 45 | 48 |
| 10 | | 0 | 3 | 5 | 8 | 11 | 14 | 17 | 20 | 23 | 26 | 29 | 33 | 36 | 39 | 42 | 45 | 48 | 52 | 55 |
| 11 | | 0 | 3 | 6 | 9 | 13 | 16 | 19 | 23 | 26 | 30 | 33 | 37 | 40 | 44 | 47 | 51 | 55 | 58 | 62 |
| 12 | | 1 | 4 | 7 | 11 | 14 | 18 | 22 | 26 | 29 | 33 | 37 | 41 | 45 | 49 | 53 | 57 | 61 | 65 | 69 |
| 13 | | 1 | 4 | 8 | 12 | 16 | 20 | 24 | 28 | 33 | 37 | 41 | 45 | 50 | 54 | 59 | 63 | 67 | 72 | 76 |
| 14 | | 1 | 5 | 9 | 13 | 17 | 22 | 26 | 31 | 36 | 40 | 45 | 50 | 55 | 59 | 64 | 67 | 74 | 78 | 83 |
| 15 | | 1 | 5 | 10 | 14 | 19 | 24 | 29 | 34 | 39 | 44 | 49 | 54 | 59 | 64 | 70 | 75 | 80 | 85 | 90 |
| 16 | | 1 | 6 | 11 | 15 | 21 | 26 | 31 | 37 | 42 | 47 | 53 | 59 | 64 | 70 | 75 | 81 | 86 | 92 | 98 |
| 17 | | 2 | 6 | 11 | 17 | 22 | 28 | 34 | 39 | 45 | 51 | 57 | 63 | 67 | 75 | 81 | 87 | 93 | 99 | 105 |
| 18 | | 2 | 7 | 12 | 18 | 24 | 30 | 36 | 42 | 48 | 55 | 61 | 67 | 74 | 80 | 86 | 93 | 99 | 106 | 112 |
| 19 | | 2 | 7 | 13 | 19 | 25 | 32 | 38 | 45 | 52 | 58 | 65 | 72 | 78 | 85 | 92 | 99 | 106 | 113 | 119 |
| 20 | | 2 | 8 | 13 | 20 | 27 | 34 | 41 | 48 | 55 | 62 | 69 | 76 | 83 | 90 | 98 | 105 | 112 | 119 | 127 |

*Table 4.5 Critical values for the Mann-Whitney U test (at the p = 0.05 level). If the smallest U value is less than or equal to the critical value then there is a significant difference between the two sets of data*

If you compare the values of U in Box 4.3 with the overlap line graphs in Figure 4.1. You will notice that as the overlap between the data increases, so the smallest *U* value gets bigger. The data on worm numbers shows very little overlap and has the lowest value of $U = 1$. The moisture content shows more overlap and has a larger value of $U = 8$. The organic content shows the greatest overlap and has the highest value of $U = 16$.

As in the *t-Test* the critical value helps to indicate a cut-off point along this scale of increasing overlap. When the smallest *U* value exceeds the critical value then there is so much overlap that it is no longer possible to be certain there is a significant difference. Similarly as in the t-Test the critical value depends on the sample size. For the moisture and organic content, values of $n_1 = 7$ and $n_2 = 6$ were used to find the critical value. This was because Sanjit lost one of his soil samples during his investigation and so only had six results for the moisture and organic content in the second field.

***Using the Mann-Whitney U Test***

- it is used to test the difference between two sets of data.
- it makes no assumptions about the distribution of the data, i.e. is *distribution free*. For this reason, it can be used with non-normal data.
- it needs at least six measurements within each data set.
- it is not suitable for sample sizes of greater than twenty measurements (use the t-Test instead).
- if *U* is *less than or equal to* the critical value then it is possible to reject the null hypothesis and accept that there *is a significant difference* between the two sets of data.

*Box 4.4*

## PRESENTING RESULTS OF STATISTICAL TESTS

Presenting the results of your tests in a table like the one shown below can simplify the interpretation of your data.

| | **Median** | | | | |
|---|---|---|---|---|---|
| | **Field 1** | **Field 2** | ***U* value** | **Critical value** | **Significance** |
| **Number of worms** | 3<br>$n_1 = 7$ | 14<br>$n_2 = 7$ | 1 | 8 | Yes<br>($p < 0.05$) |
| **Moisture content/%** | 41<br>$n_1 = 7$ | 35<br>$n_2 = 6$ | 8 | 6 | No<br>($p > 0.05$) |
| **Organic content/%** | 35<br>$n_1 = 7$ | 31.5<br>$n_2 = 7$ | 16 | 6 | No<br>($p > 0.05$) |

*Table 4.6 A summary of Sanjit's data*

So far, we have looked at two methods of comparing sets of data, the *t-Test* and the *Mann-Whitney U test*. These tests are designed for use with *independent* data. In other words, the value of a measurement in one data set is not linked in any way to a value in the other set. There are occasions, however, when it is necessary to test for differences between data sets which are in some way linked or *dependent*. The *Wilcoxon matched pairs test* is designed for these occasions.

## WILCOXON MATCHED PAIRS TEST - A WORKED EXAMPLE

### Case study 6

Siobhan and Leanne were carrying out an investigation to compare the amount of mosses found on opposite sides of oak trees. They knew that mosses were dependent on moisture and had an idea that there would be a greater percentage cover of mosses on the shaded, north-facing sides of trees than on the sunny, south-facing sides.

They found the percentage cover of mosses by recording the presence or absence at 100 points on a 50 cm x 50 cm gridded quadrat. The quadrat was placed at a height of 1m on both the north and south sides of ten oak trees. Their data are shown in Table 4.7.

Initially they used the *Mann Whitney U Test* to compare their data. However, they found that there was so much variation between the ten trees that it completely masked any difference between the north and south sides of the trees.

The median percentage covers values were 35.5 for the north and 33 for the south.

In any wood there will be some trees growing in bright open glades and others growing in dense shaded regions. The variation between trees is therefore huge. In Siobhan and Leanne's data, however, there still seemed to be an almost constant pattern of more moss cover on the north side of the trees than on the south side. The data also formed *matched pairs*, i.e. a tree with lots of moss on the south side was very likely to have a lot of moss on the north side and vice versa. They decided that as the data seemed to be matched they would use the *Wilcoxon matched pairs test*.

Siobhan and Leanne started with the null hypothesis that there is no difference between the moss cover on the north side and south side of the trees.

| Tree | Growth on North side/% $x_1$ | Growth on South side/% $x_2$ | Difference $x_1-x_2$ | Rank | Sign |
|---|---|---|---|---|---|
| 1 | 12 | 9 | 3 | 3 | + |
| 2 | 35 | 34 | 1 | 1 | + |
| 3 | 42 | 36 | 6 | 5 | + |
| 4 | 29 | 27 | 2 | 2 | + |
| 5 | 15 | 7 | 8 | 6 | + |
| 6 | 74 | 65 | 9 | 7 | + |
| 7 | 43 | 19 | 24 | 8 | + |
| 8 | 36 | 36 | 0 | – | |
| 9 | 27 | 32 | –5 | 4 | – |
| 10 | 29 | | 27 | 9 | + |

*Table 4.7 Siobhan and Leanne's results on percentage cover of moss on a sample of ten trees*

*Step 1* Calculate the difference between each pair of measurements ($x_1 - x_2$).

*Step 2* Rank the differences ignoring the signs. (Zero does not get a rank value.)

*Step 3* Note where the difference was positive or negative and put a + or – sign against each corresponding rank value

*Step 4* Sum the positive and negative rank values (separately).

$$\Sigma R \text{ positive} = 41$$
$$\Sigma R \text{ negative} = 4$$

*Step 5* Compare the smaller $\Sigma R$ value in step 4 with the critical value for the appropriate degrees of freedom (see Table 4.8).

The degrees of freedom is taken as the number of non-zero differences. If the smaller $\Sigma R$ value is *less than or equal to* the critical value then there is a significant difference between the two sets of data.

For 9 degrees of freedom the critical value = 5.

Siobhan and Leanne's smaller $\Sigma R$ value (i.e. 4) is less than the critical value (i.e. 5), therefore there is a significant difference between the moss cover on the north and south side of the trees (at the $p = 0.05$ level).

The null hypothesis is rejected ($p < 0.05$).

| Degrees of freedom | Critical value |
|---|---|
| 6 | 0 |
| 7 | 2 |
| 8 | 3 |
| 9 | 5 |
| 10 | 8 |
| 11 | 10 |
| 12 | 13 |
| 13 | 17 |
| 14 | 21 |
| 15 | 25 |
| 16 | 29 |
| 17 | 34 |
| 18 | 40 |
| 19 | 46 |
| 20 | 52 |
| 21 | 58 |
| 22 | 65 |
| 23 | 73 |
| 24 | 81 |
| 25 | 89 |
| 26 | 98 |
| 27 | 107 |
| 28 | 116 |
| 29 | 126 |
| 30 | 137 |

*Table 4.8 Critical values for the Wilcoxon Matched Pairs Test (at the $p = 0.05$ level). If $\Sigma R$ is less than or equal to critical value then there is a significant difference between the two sets of data. Degrees of freedom are given by the number of non-zero differences*

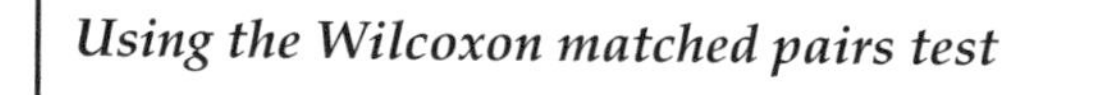

***Using the Wilcoxon matched pairs test***

- it is only suitable for comparing data which has been collected in *matched pairs*.
- it makes no assumption about the distribution of the data and so can be used for non-normal data.
- it needs at least six pairs of measurements. In practical terms it should not be used with more than thirty pairs.
- if $\Sigma R$ is *less than or equal to* the critical value it is possible to reject the null hypothesis and accept that there *is a significant difference* between the two sets of data.

*Box 4.5*

# Chapter 5 Correlations and associations

## Case study 7

*Carl sampling in a stream*

Carl decided to carry out an investigation to look at the effect of flow rate on the density of stream invertebrates. His first thought was to choose both a fast-flowing section and a slow-flowing section of the stream and to collect invertebrates by sampling within six randomly placed quadrats in each section. He would then compare the median number of invertebrates found for each site using the *Mann-Whitney U Test*.

However, when Carl got to the stream he found that there were no discrete fast and slow-flowing sections that were large enough to allow him to take six samples. The stream, in fact, consisted of a variety of short sections ranging in flow from torrential waterfalls to slow pools. Carl decided to choose twelve sites at random and using a quadrat at each site he measured the flow rate and density of invertebrates.

The data collected by Carl is shown in Table 5.1. Looking at the data it seemed as if there were more invertebrates in the fast-flowing sections than in the slow-flowing sections. Carl plotted a scattergram to illustrate his results (see Fig.5.1).

| Site | Flow/ms$^{-1}$ | Invertebrate density/ number per 0.5 × 0.5 m quadrat |
|---|---|---|
| 1 | 0.3 | 86 |
| 2 | 0.6 | 46 |
| 3 | 0.5 | 39 |
| 4 | 0.1 | 15 |
| 5 | 0.1 | 41 |
| 6 | 0.4 | 52 |
| 7 | 0.9 | 100 |
| 8 | 0.8 | 63 |
| 9 | 0.2 | 60 |
| 10 | 0.5 | 30 |
| 11 | 0.7 | 72 |
| 12 | 1.1 | 71 |

*Table 5.1 Carl's results from randomly chosen sites along the stream.*

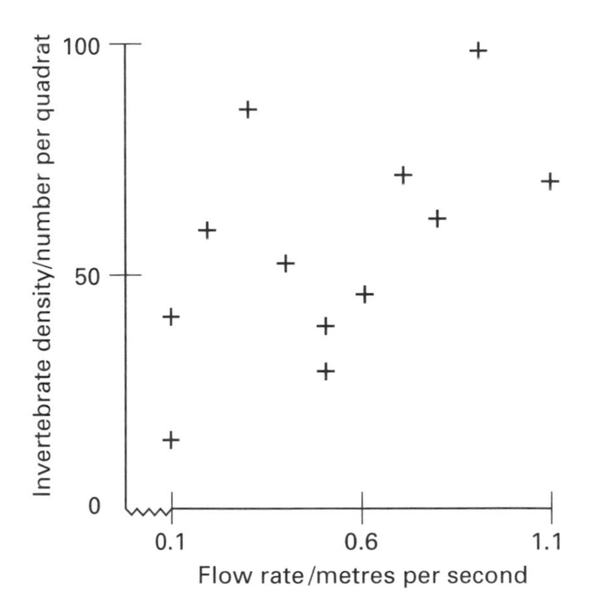


*Figure 5.1 Relationship between flow rate and invertebrate density*

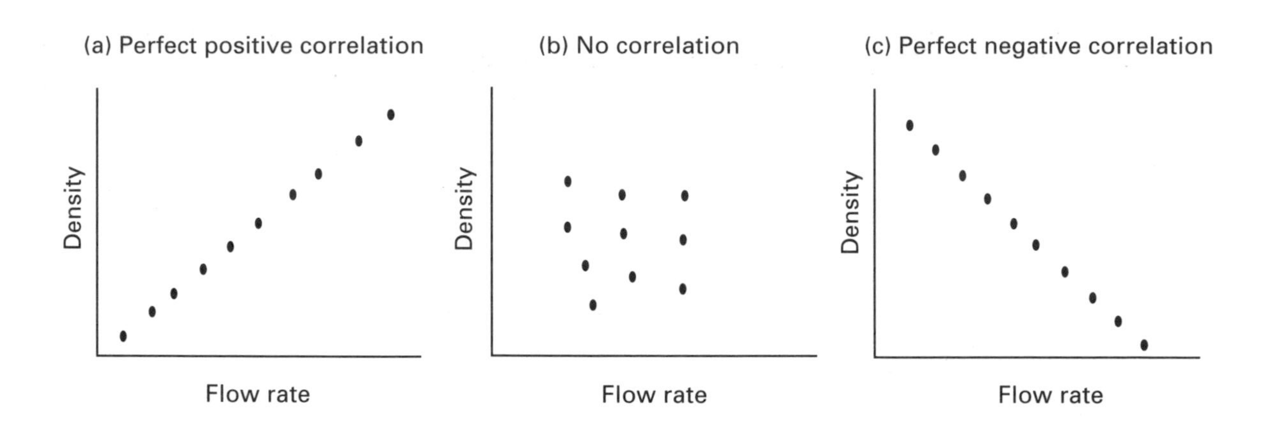


*Figure 5.2 Compare Carl's scattergram with the three diagrams above. In graph (a) for every increase in flow rate there is an increase in the number of animals, i.e. there is a perfect positive correlation. In graph (c) for every increase in flow rate there is a decrease in the number of animals, i.e. there is a perfect negative correlation. In graph (b) there is a random scattering of points, i.e. there is no correlation.*

Looking at the scattergram Carl decided that although the points were quite scattered there appeared to be some sort of correlation between the flow rate and the density of invertebrates. It wasn't a perfect correlation though (compare Carl's scattergram with Fig.5.2(a)). Carl realised that he needed a statistical test to tell him whether there was a sufficient likelihood that this was a true correlation and not merely due to chance.

In a set of data like those of Carl's shown in Table 5.1, it is possible to *rank* the flow rates and the number of animals. If there is a perfect positive correlation then each pair of measurements should have the same rank for flow rate and for number of animals. We can express this mathematically by calculating the difference ($D$) between the ranks.

For a perfect positive correlation $D$ will always be zero. Any deviation from this perfect situation will produce positive and negative differences between the ranks. The largest values of $D$ (ignoring any signs) will be found when there is a perfect negative correlation (see Fig.5.2(c)).

Carl thought it might be an idea to add all the $D$ values to give a measure of correlation. Unfortunately, you will find that the –ve $D$ values cancel out the +ve $D$ values. This problem can be overcome by squaring the differences. The sum of these squared values ($\sum D^2$) gives a measure of the correlation between two variables. If the data are close to a perfect +ve correlation then $\sum D^2$ will be very small, if they are close to a perfect negative correlation then $\sum D^2$ will be large.

For convenience, the measure of correlation lies between the values of +1 (indicating a perfect positive correlation) and –1 (a perfect negative correlation). This measure of correlation is called the *Spearman Rank Correlation Coefficient* ($r_s$) and is found by using the following formula:

$$r_s = 1 - \frac{6\sum D^2}{n(n^2 - 1)}$$

| Site | Flowrate/$ms^{-1}$ | Rank $R_1$ | Animals per quadrat | Rank $R_2$ | $D$ ($R_1$–$R_2$) | $D^2$ |
|---|---|---|---|---|---|---|
| 1 | 0.3 | **4** | 86 | **11** | –7 | 49 |
| 2 | 0.6 | **8** | 46 | **5** | +3 | 9 |
| 3 | 0.5 | **6.5** | 39 | **3** | +3.5 | 12.25 |
| 4 | 0.1 | **1.5** | 15 | **1** | +0.5 | 0.25 |
| 5 | 0.1 | **1.5** | 41 | **4** | –2.5 | 6.25 |
| 6 | 0.4 | **5** | 52 | **6** | –1 | 1 |
| 7 | 0.9 | **11** | 100 | **12** | –1 | 1 |
| 8 | 0.8 | **10** | 63 | **8** | +2 | 4 |
| 9 | 0.2 | **3** | 60 | **7** | –4 | 16 |
| 10 | 0.5 | **6.5** | 30 | **2** | +4.5 | 20.25 |
| 11 | 0.7 | **9** | 72 | **10** | –1 | 1 |
| 12 | 1.1 | **12** | 71 | **9** | +3 | 9 |
| | | | | Σ(Sum) | 0 | 129 |

*Table 5.2 Rank values and differences for Carl's results*

*Step 1* Rank the data for each variable (see Table 5.1).

*Step 2* Calculate the difference ($D$) between each pair of ranks. (If your calculations are correct then the sum of the differences should equal zero.)

*Step 3* Square the differences ($D^2$)

*Step 4* Sum the $D^2$ values. $\Sigma D^2 = 129$

*Step 5* Calculate $r_s$.

$$r_s = 1 - \frac{6\Sigma D^2}{n(n^2 - 1)}$$

$$= 1 - \frac{6 \times 129}{12 \times 143}$$

$= 1 - 0.45 \quad r_s = 0.55$

*Step 6* Compare the value of $r_s$ against the critical value for the appropriate number of pairs of measurements ($n$).

For $n = 12$ the critical value = 0.59.

| Number of pairs of measurements | Critical value |
|---|---|
| 5 | 1.00 |
| 6 | 0.89 |
| 7 | 0.79 |
| 8 | 0.74 |
| 9 | 0.68 |
| 10 | 0.65 |
| 12 | 0.59 |
| 14 | 0.54 |
| 16 | 0.51 |
| 18 | 0.48 |
| 20 | 0.45 |
| 22 | 0.43 |
| 24 | 0.41 |
| 26 | 0.39 |
| 28 | 0.38 |
| 30 | 0.36 |

*Table 5.3 Critical values for the Spearman Rank Correlation (at the p = 0.05 level). If $r_s$ (ignoring the sign) is greater than or equal to the critical value for the appropriate value of n then there is a significant correlation. If $r_s$ is +ve there is a positive correlation. If $r_s$ is –ve there is a negative correlation.*

Carl's value of $r_s$ (i.e. 0.55) is less than the critical value (i.e. 0.59) therefore there is no significant correlation between flow rate and number of animals.

Carl was a little disappointed at this result. However, he soon realised that simply grouping all the animals together to give a total number was not very satisfactory. Animals such as blackfly larva are filter feeders that depend on flowing water to bring them food. Others, such as freshwater shrimps, are deposit feeders and prefer still water.

Carl decided to analyse his data a second time using the numbers of different groups present within each quadrat. His results are shown in Table 5.4. Look at the table and see if you can work out the same answers as those given here.

After further analysis, Carl's results show that for both blackflies and shrimps the calculated value of $r_s$ (ignoring the sign) is greater than the critical value. Therefore, there is a significant positive correlation between flow rate and the number of blackfly larvae per quadrat and a significant negative correlation between flow rate and the number of shrimps per quadrat.

A significiant correlation does not necessarily mean that changes in one factor cause changes in the other factor; there could be some third factor responsible for changes in both. It is sometimes possible to devise an experimental test to help you decide if there is any form of causative relationship between two variables. Ideally you should try to control all variables except the one that you think is important.

***Using the Spearman rank correlation:***

- if $r_s$ (ignoring any sign) is *greater than or equal to* the critical value then there *is a significant correlation* between the two variables. A positive value of $r_s$ indicates a positive correlation and a negative value a negative correlation.

- a minimum of five pairs of measurements are needed. In practice ten to fifteen pairs give a better chance of finding a significant correlation.

*Box 5.1*

| Site | Flow /ms$^{-1}$ | Rank $R_1$ | Blackflies per quadrat | Rank $R_2$ | $D$ $(R_1 - R_2)$ | $D^2$ | Shrimps per quadrat | Rank $R_3$ | $D$ $(R_1 - R_3)$ | $D^2$ |
|---|---|---|---|---|---|---|---|---|---|---|
| 1 | 0.3 | **4** | 10 | **5** | | | 23 | **10** | | |
| 2 | 0.6 | **8** | 31 | **8** | | | 2 | **5.5** | | |
| 3 | 0.5 | **6.5** | 26 | **7** | | | 0 | **2** | | |
| 4 | 0.1 | **1.5** | 1 | **2** | | | 10 | **8** | | |
| 5 | 0.1 | **1.5** | 0 | **1** | | | 35 | **12** | | |
| 6 | 0.4 | **5** | 8 | **4** | | | 11 | **9** | | |
| 7 | 0.9 | **11** | 73 | **12** | | | 0 | **2** | | |
| 8 | 0.8 | **10** | 54 | **10** | | | 1 | **4** | | |
| 9 | 0.2 | **3** | 5 | **3** | | | 34 | **11** | | |
| 10 | 0.5 | **6.5** | 21 | **6** | | | 2 | **5.5** | | |
| 11 | 0.7 | **9** | 42 | **9** | | | 9 | **7** | | |
| 12 | 1.1 | **12** | 65 | **11** | | | 0 | **2** | | |
| | | | $n$ = 12<br>critical value = 0.59 | | $\Sigma D^2$ = 5<br>$r_s$ = 0.98 | | $n$ = 12<br>critical value = 0.59 | | $\Sigma D^2$ = 517<br>$r_s$ = –0.81 | |

*Table 5.4 Further analysis of Carl's stream data.*

# Chapter 6 Dealing with categorical data

So far, we have looked at data that have been *interval* measurements; things like height, leaf width, insect density, flow rate and light intensity. The important characteristic of these data is that it is possible to say just how much bigger one measurement is than another, for example, a leaf 15 cm wide is three times bigger than a leaf 5 cm wide. In other words, interval measurements are *real* measurements.

Sometimes it is only possible to say that data are *categorical*, i.e. that they belong to one of a number of categories.

### Case study 8

Fiona and Sangeeta were carrying out an investigation to look at the distribution of moorland plants. They had an idea that bilberry (*Vaccinium myrtillus*) and ling (*Calluna vulgaris*) might be *associated*. In other words, if bilberry was present within an area then there was a good chance that ling was also present. To test this idea they looked at the presence and absence of the two species within 200 randomly placed quadrats. Their results are shown in Table 6.1.

| | **Ling present** | **Ling absent** |
|---|---|---|
| **Bilberry present** | 74 | 51 |
| **Bilberry absent** | 18 | 57 |

*Table 6.1 Fiona and Sangeeta's categorical data*

Fiona and Sangeeta found that 74 quadrats contained both ling and bilberry. They also calculated the number of quadrats they would have expected to contain both species (see Box 6.1). They found that if the two species were randomly distributed they would have expected 57.5 quadrats to contain both species.

***Calculating observed and expected values***

Total number of quadrats containing ling:

Column total = 74 + 18 = 92

Total number of quadrats containing bilberry:

Row total = 74 + 51 = 125

Total number of quadrats examined

Grand total = 200

To calculate the expected value for both species we use the formula:

$$E = \frac{(\text{Row total}) \times (\text{Column total})}{\text{Grand total}}$$

Expected number of quadrats containing both ling and bilberry

$$E = \frac{125 \times 92}{200} = 57.5$$

Observed number of quadrats containing both ling and bilberry from Table 6.1

$$O = 74$$

*Box 6.1*

Fiona and Sangeeta needed to determine whether or not their observed value of 74 was significantly different from their expected value of 57.5.

There is a simple statistical test which looks at the difference between *observed* and *expected* values and relates them to a probability level, thus making it possible to identify how likely it is that the values are significantly different. The test is called the *Chi squared test*.

### *The Chi squared test*

The formula for the Chi squared test is as follows:

$$\chi^2 = \Sigma \frac{(O-E)^2}{E}$$

where $\chi$ is the Greek letter Chi
$O$ is the *observed* value
$E$ is the *expected* value.

The top part of the formula for $\chi^2$ considers the size of the difference between the observed and expected values. This difference could be either positive or negative. To avoid the mathematical problems associated with negative values, the difference is squared.

The bottom part of the formula relates the size of the difference to the magnitude of the numbers involved. The sigma ( ) 'sum' symbol is required because there is not just one pair of observed and expected values, but several (in this case four).

Fiona and Sangeeta calculated the expected value for the number of quadrats containing both species (see Box 6.1). They also worked out expected values for the other categories using the same method. Their results are shown in Table 6.2.

| | **Ling present** | **Ling absent** |
|---|---|---|
| **Bilberry present** | $O = 74$<br>$E = 57.5$ | $O = 51$<br>$E = 67.5$ |
| **Bilberry absent** | $O = 18$<br>$E = 34.5$ | $O = 57$<br>$E = 40.5$ |

*Table 6.2 Fiona and Sangeeta's observed and expected values*

By taking all of the observed and expected values from Table 6.2 and putting them into the formula, Fiona and Sangeeta calculated $\chi^2$.

$$\chi^2 = \Sigma \frac{(O-E)^2}{E}$$

$$= \frac{(74-57.5)^2}{57.5} + \frac{(18-34.5)^2}{34.5} + \frac{(51-67.5)^2}{67.5} + \frac{(57-40.5)^2}{40.5}$$

$$= 4.73 + 7.89 + 4.03 + 6.72$$

$$\chi^2 = 23.37$$

A large value of $\chi^2$ occurs when there is a big difference between observed and expected values. So the larger the $\chi^2$ value, the more certain it is that the difference is significant. As with all the other tests it is necessary to consult a table of critical values to pick a cut-off point (see Table 6.3). The degrees of freedom to be used can be found as follows:

(No. of rows – 1) × (No. of columns –1)

As there are two rows and two columns in Fiona and Sangeeta's results (see Table 6.2), the degrees of freedom are:.

$(2-1) \times (2-1) = 1$

As Fiona and Sangeeta's calculated value of $\chi^2$ (i.e. 23.37) is greater than the critical value (i.e. 3.84) then there is a significant difference between the observed and expected values.

As the observed value is *greater than* the expected value it is possible to say that ling and bilberry are *associated*, i.e. they grow together more often than could be expcted by chance. (If the observed value had been *less than* the expected then the two species could be said to be *dissociated*, growing together less often than would be expected.)

The null hypothesis is rejected ($p < 0.05$).

### Case study 9

Mark and Paula carried out an investigation to study what happens when woodlice are given a choice between dry and humid atmospheres. Their investigation consisted of five trials with ten animals used in each trial. Their results are shown in Table 6.4.

| Trial | Distribution of ten woodlice after three minutes | |
|---|---|---|
| | Dry atmosphere | Humid atmosphere |
| 1 | 3 | 7 |
| 2 | 4 | 6 |
| 3 | 3 | 7 |
| 4 | 5 | 5 |
| 5 | 4 | 6 |
| Total | $O = 19$ | $O = 31$ |

*Table 6.4 Mark and Paula's data*

Mark and Paula wanted to find out if their results showed a significant difference in the distribution of the woodlice or was it due to pure chance. They started with the null hypothesis that there is no difference between the numbers of woodlice found in dry and humid conditions. If this was true they would expect to find 25 animals in each area (i.e. half of the total).

Putting the observed and expected values into the Chi squared formula:

$$\chi^2 = \Sigma \frac{(O-E)^2}{E} \qquad \chi^2 = 2.88$$

Degrees of freedom = number of categories minus one. In this case $2 - 1 = 1$.

The critical value at one degree of freedom is 3.84 (at the $p = 0.05$ level).

| Degrees of freedom | Critical value |
|---|---|
| 1 | 3.84 |
| 2 | 5.99 |
| 3 | 7.81 |
| 4 | 9.49 |
| 5 | 11.07 |
| 6 | 12.59 |
| 7 | 14.07 |
| 8 | 15.51 |
| 9 | 16.92 |
| 10 | 18.31 |
| 11 | 19.68 |
| 12 | 21.02 |
| 13 | 22.36 |
| 14 | 23.69 |
| 15 | 24.99 |
| 16 | 26.30 |
| 17 | 27.59 |
| 18 | 28.87 |
| 19 | 30.14 |
| 20 | 31.41 |
| 21 | 32.67 |
| 22 | 33.92 |
| 23 | 35.17 |
| 24 | 36.42 |
| 25 | 37.65 |
| 26 | 38.89 |
| 27 | 40.11 |
| 28 | 41.34 |
| 29 | 42.56 |
| 30 | 43.77 |

*Table 6.3 Critical values for the Chi squared test (at the $p = 0.05$ level). If $\chi^2$ is greater than or equal to the critical value then there is a significant difference between the observed and expected values.*

Mark and Paula's value of $\chi^2$ (i.e. 2.88) is less than the critical value (i.e. 3.84) therefore, there is no significant difference between the observed and expected values.

The null hypothesis is accepted ($p > 0.05$).

*Yates' correction*

To be strictly accurate when dealing with only two categories, we should modify the $\chi^2$ equation by subtracting 0.5 from each value of $|O - E|$. This is called the *Yates correction* and gives a modified formula:

$$\chi^2 = \Sigma \frac{(|O - E| - 0.5)^2}{E}$$

In Mark and Paula's investigation, the modified value of $\chi^2$ would be 2.42. This is still less than the critical value so the null hypothesis is still accepted.

Box 6.2

## GENETICS AND GREGOR MENDEL

*Postage stamp showing Gregor Mendel and the plants he worked with*

In one of his famous experiments, Mendel crossed tall pea plants with short pea plants. He then crossed the $F_1$ offspring and observed 787 tall plants and 277 short ones in the resultant $F_2$ offspring. He suggested that this was a true 3 : 1 ratio of tall:short.

The genetics ratio of 3:1 is possibly familiar to you and the breeding experiments where you would find offspring in this ratio. In real life, however, breeding sweet pea plants are seldom so mathematically obliging and off-spring are rarely found in exact ratios.

In the example above the total number of offspring is 1064 plants. If the ratio was exactly 3:1 Mendel would have expected 798 tall plants and 266 short plants in the resultant $F_2$ offspring. How close was his observed ratio to the expected ratio?

| | | |
|---|---|---|
| Tall plants | $O = 787$ | $E = 798$ |
| Short plants | $O = 277$ | $E = 266$ |

Putting the observed and expected values into the modified Chi squared test:

$$\chi^2 = \Sigma \frac{(|O - E| - 0.5)^2}{E}$$

$$= \frac{(|787 - 798| - 0.5)^2}{798} + \frac{(|277 - 266| - 0.5)^2}{266}$$

$$= 0.14 + 0.41$$

$$\chi^2 = 0.55$$

Degrees of freedom = 1
(i.e. number of categories minus one)

The critical value at one degree of freedom is 3.84 (at the p = 0.05 level). Mendel's $\chi^2$ value (i.e. 0.55) is well below the critical value (i.e. 3.84) therefore there is no significant difference between the observed and expected values. Mendel's results, therefore, can be accepted as a true 3:1 ratio.

# Chapter 7 Other facts and formulae

There are other aspects of mathematics books that are outside the scope of this book. We assume that you will already have acquired skills in basic number work and in using a calculator. You may need to find out more about graphs (including the 'area under a graph' and working out rates of reaction.)

In this final chapter, we include some further useful information and strongly advise you to work through all the Examination Questions at the end of the book.

## SIMPSON'S DIVERSITY INDEX

The measure of *species richness* relates to the number of different species found in a habitat. Old grassland meadows, for example, are described as being species rich and are very attractive to the naturalist. In a list indicating 'richness' the same weight is given to a single specimen of one species as to the dominant species which could be present in hundreds of thousands.

A more accurate measure of the number and abundance of species is found using the *Simpson's Diversity Index* ($D$).

$$D = \frac{N(N-1)}{\sum n(n-1)}$$

There are other types of diversity index, but this method is simple and is a useful means of comparing different areas or subsets of a total population.

A high value of $D$ indicates a stable and ancient site, such as a species-rich meadow or an old woodland. A low value of $D$ may suggest agricultural management, a recently colonised site or a polluted area. These are only generalisations and suggest that perhaps a further investigation would be worthwhile. (The meaning of $N$ and $n$ is explained in the worked example.)

### A worked example

Imagine you have looked at the whole flower population of a meadow by sampling in a random way (e.g. using point quadrats).

You should record your results as shown in Table 7.1. If you are unable to name a particular species, give it a 'letter' code and then you can recognise it each time you find it.

| **Species** | **Number** $n$ | $n(n-1)$ |
|---|---|---|
| Festuca | 54 | $54 \times 53 = 2862$ |
| Buttercup | 17 | $17 \times 16 = 272$ |
| Shepherd's purse | 12 | $12 \times 11 = 132$ |
| Species 'A' | 5 | $5 \times 4 = 20$ |
| Dandelion | 3 | $3 \times 2 = 6$ |
| Total (N) | 91 | 3292 |

*Table 7.1 Flower population from a field*

Putting the results from the table into the Simpson's diversity index equation:

$$D = \frac{N(N-1)}{\sum n(n-1)}$$

$$D = \frac{91 \times 90}{3292} = \frac{8190}{3292} = 2.49$$

## LINCOLN INDEX

When a species is restricted to a certain area (e.g. diving beetles in a pond or snails in a walled garden) and is studied, it is possible to estimate population numbers by a capture, marking and recapture technique. This method (known as *population sampling*) involves taking a random sample of the population, (i.e. by netting or using mammal traps), marking the trapped individuals in such a way that they are neither damaged nor made more conspicuous to predators and then releasing them.

After a suitable time, to allow the original sample to mingle with the population again, another trapping exercise is carried out and the proportion of previously marked individuals is calculated. This proportion in the second sample is assumed to be the same as the proportion of marked individuals in the whole population. Since the total number of marked specimens is known it is possible to estimate by ratio the size of the population living in the area being studied.

The formula for the *Lincoln Index* is as follows:

$$\frac{\text{Number marked in 2nd sample}}{\text{Total caught in 2nd sample}} = \frac{\text{Number marked in first sample}}{\text{Size of whole population } (n)}$$

### *A worked example*

Imagine you laid some traps to catch mice in a warehouse. You trap eighteen mice and mark them by clipping fur from the chest. You release them and on the next night you set the traps again. This time you trap twenty four mice, of which eight were found to have been marked from the previous night.

Putting these results into the Lincoln Index equation, we find that

$$\frac{8}{24} = \frac{18}{n} \quad \text{i.e.} \quad n = \frac{18 \times 24}{8} = 54.$$

## POWERS OF TEN

In scientific notation *powers of ten* are used to avoid writing numerous zeros to indicate very large or very small numbers. The first bacteria on Earth existed three thousand five hundred million years ago, this can be written as 3 500 000 000, or more simply as $3.5 \times 10^9$ years. If one of those bacteria happened to be 1.7 micrometres in diameter (a micrometre (µm) is one millionth of a metre) this diameter could be written as 0.0000017m, or more simply as $1.7 \times 10^{-6}$m.

| Prefix | Factor | Sign |
|---|---|---|
| kilo– | $\times 10^3$ | k |
| mega– | $\times 10^6$ | M |
| giga– | $\times 10^9$ | G |
| milli– | $\times 10^{-3}$ | m |
| micro– | $\times 10^{-6}$ | µ |
| nano– | $\times 10^{-9}$ | n |

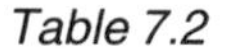
*Table 7.2*

## SI UNITS

| | Measurement | Unit | Symbol |
|---|---|---|---|
| (a) | *Basic units* | | |
| | Length | metre | m |
| | Mass | kilogramme | kg |
| | Time | second | s |
| | Amount of substance | mole | mol |
| (b) | *Derived units* | | |
| | Force | newton | N |
| | Energy | joule | J |
| | Pressure | pascal | Pa |

*Table 7.3*

## SPECIES FREQUENCY AND ABUNDANCE

*Species frequency* can be measured by recording the percentage of random quadrats in which a particular species is found.

$$\%\ \text{presence} = \frac{\text{number of quadrats containing the species}}{\text{total number of quadrats}} \times 100$$

*Species abundance* can be determined by dividing a quadrat into 25 squares and estimating how many of these are covered by the species in question. Each small square is 4% of the total area so, for example, a species that is present in 8 squares covers 32%.

A *subjective estimate of relative abundance* is often recorded as the *ACFOR scale*. In this technique you simply write against the species list your estimate of whether it is:

**A**bundant,
**C**ommon,
**F**requent,
**O**ccasional,
**R**are
(hence **ACFOR**).

This scale can be used for plants or animals (only those that are fixed to one spot).

Sometimes investigators use the additional classes, Super-abundant (S) and Extremely-abundant (E). The following example (Table 7.4) is taken from *The OU Project Guide*, Field Studies Council, 1986 (modified from the original by Crisp and Southward, 1958).

| Symbol | Algae | Limpets |
|---|---|---|
| E | > 90% cover | 20+ per 0.1 $m^2$ |
| S | 60-89% | 10-19 per 0.1 $m^2$ |
| A | 30-50% | 5-9 per 0.1 $m^2$ |
| C | 5-29% | 1-4 per 0.1 $m^2$ |
| F | < 5% (zone distinct) | 5-9 per $m^2$ |
| O | scattered (zone not distinct) | 1-4 per $m^2$ |
| R | only 1-2 plants | < 1 per $m^2$ |

*Table 7.4*

This technique is used to calculate frequencies of genotypes and alleles in populations. The two mathematicians after whom the equation is named made the assumption that, under certain conditions, genotype frequencies would remain the same for generation after generation.

Imagine the case where there are two alleles, **A** and **a**, for a particular genetic condition. Each diploid individual could be **AA**, **Aa** or **aa**. Suppose that the allele **A** is present in the population at a particular frequency, say 20% (this could also be represented as 20/100 or as 0.2). Since together **A** and **a** must add up to 100%, then allele **a** must be 80% (or 80/100 or 0.8) of the population.

Hardy and Weinberg represented the frequencies of the two alleles as '$p$' and '$q$'. So the frequency of allele **A** is $p = 0.2$ and the frequency of allele **a** is $q = 0.8$ and, as we have seen, $p + q = 1$, i.e. $0.2 + 0.8 = 1$.

Table 7.5 shows the possible combinations of alleles to give the different genotypes. The Hardy Weinberg representations of these frequencies are shown in the brackets.

| | | Male gametes and frequency | |
|---|---|---|---|
| | | **A** ($p$) | **a** ($q$) |
| Female gametes and frequency | **A** ($p$) | **AA** ($p^2$) | **Aa** ($pq$) |
| | **a** ($q$) | **Aa** ($pq$) | **aa** ($q^2$) |

*Table 7.5*

The total population adds up to 100% (for equilibrium equations this is shown as 1). Therefore, by adding up all of the combinations in Table 7.5, we should get a value of 1.

$$p^2 + 2pq + q^2 = 1$$

If $p + q = 1$

then $(p + q)^2 = 1$

Therefore $p^2 + 2pq + q^2 = 1$.

Make sure that you understand this technique by looking at the worked example below.

(For the further information on the genetics presented here, you could refer to *Biology Advanced Studies - Genetics and Evolution*, Nelson.)

### *A worked example*

In one animal species hairlessness is caused by a recessive allele **h** and 25% of the population are hairless. What is the frequency of the heterozygous and homozygous individuals?

*Step 1* The hairless genotype **hh** has a frequency of 25%.

so

$$\mathbf{hh} = q^2 = 0.25$$

therefore

$$q = \sqrt{0.25} = 0.5$$

*Step 2* As $p + q = 1$, so $p = 1 - q = 1 - 0.5$

therefore $p = 0.5$

*Step 3* Calculate the value of $p^2$ (to find **HH**).

$$p^2 = 0.5^2 = 0.25 \text{ or } 25\%.$$

The frequency of the homozygous **HH** is 25%.

*Step 4* Calculate the value of $2pq$ (to find **Hh**).

$$2pq = 2 \times 0.5 \times 0.5 = 0.5$$

The frequency of the heterozygote **Hh** is 50%.

## LOGARITHMIC GRAPHS

You may be familiar with the 'J-shaped' graph which is often reproduced as the population growth curve. You can make a model of this graph by calculating bacterial growth. Imagine a (culture with 125 bacteria per cm$^3$) that doubles in number every hour. Try to plot a graph of this growth.

A graph of this growth rate is shown in Figure 7.1(b). If the results of this growth rate are plotted on semi-log paper then a graph like Figure 7.1(d) is obtained. (Log-graph paper is printed so that successive intervals that appear equal are actually increasing by a factor of ten.)

(a) Table of raw data - Population numbers of the grain mite (*Acarus* sp.)

| Time/days | Number of mites in culture |
|---|---|
| 0 | 118 |
| 5 | 258 |
| 10 | 900 |
| 15 | 2190 |

(c) Part of a sheet of semi-log paper (decimal ruling on *x*-axis, logarithmic ruling on *y*-axis)

10 000 (Log 10 000 = 4) 4
1000 (Log 1000 = 3) 3
100 (Log 100 = 2) 2
10 (Log 10 = 1) 1
1 0
0 5 10 15

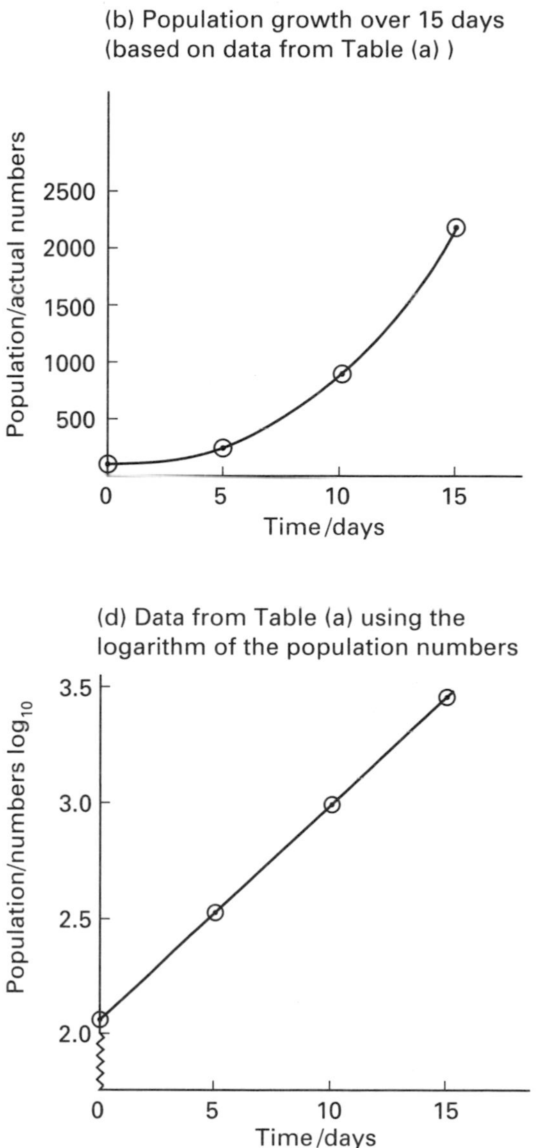


*Figure 7.1 Population increase of the grain mite Acarus sp. in culture at 25°C.*
*(Data from Soloman, Population Dynamics, IOB Studies in Biology 18, Edward Arnold)*

1. The table gives raw data of pulse rates of 25 students.

| Pulse rate of 25 students/beats min$^{-1}$ | | | | |
|---|---|---|---|---|
| 70 | 82 | 87 | 75 | 77 |
| 83 | 88 | 66 | 80 | 92 |
| 78 | 96 | 85 | 71 | 82 |
| 76 | 79 | 81 | 72 | 81 |
| 64 | 80 | 90 | 78 | 74 |

(a) Copy and complete the table below. The third line has been done for you.

| Pulse rate/ beats min$^{-1}$ | Tally | Number of students |
|---|---|---|
| 60 - 64 | | |
| 65 - 69 | | |
| 70 - 74 | //// | 4 |
| 75 - 79 | | |
| 80 - 84 | | |
| 85 - 89 | | |
| 90 - 94 | | |
| 95 - 99 | | |

(b) (i) Construct a histogram of the data in your table. (5 marks)

(ii) What is the modal class in your histogram? (1 mark)

*(AEB Human Biology, 1989, Paper 1)*

2. The activity rate of an enzyme found in human serum was measured in blood taken from a random sample of healthy adults. The following results were obtained:

| Rate of enzyme activity in arbitrary units ($x$) | Number of individuals ($f$) | ($fx$) |
|---|---|---|
| 1 | 0 | |
| 2 | 3 | |
| 3 | 2 | |
| 4 | 13 | |
| 5 | 16 | |
| 6 | 23 | |
| 7 | 18 | |
| 8 | 12 | |
| 9 | 10 | |
| 10 | 3 | |
| TOTALS | $n$ = | $fx$ = |

(a) Complete the table and then use the formula to calculate the mean value of enzyme activity for this sample. (3 marks)

(b) Represent the results given in the table in the form of a histogram. (4 marks)

(c) (i) The standard deviation for this sample is 1.8 arbitrary units. Determine the values of enzyme activity which are 2.0 standard deviation above and below the mean.

(ii) State the total number of individuals in the sample with a blood-enzyme activity value lying outside these limits. (3 marks)

(d) In the normally distributed population from which the sample was drawn, the probability (p) of an individual having a blood-enzyme activity value greater or less than 2.0 standard deviations from the mean is 0.05. What percentage of the population does this represent? (1 mark)

*(WJEC A-Level, 1991, Paper 2)*

3. In an experiment designed to test the effects of a new hormonal weedkiller, 20 barley seedlings (Batch 1) were treated with a dilute solution of the test substance one week after they had germinated. A similar batch of seedlings (Batch 2) was treated at the same time with an equal volume of water. The heights of all seedlings were measured 5 days later and the number of dead seedlings counted after two weeks. The data are presented below.

| Lengths of seedlings 5 days after treatment. Seedlings which had died after two weeks are indicated with an asterisk (*). | | | |
|---|---|---|---|
| Batch 1 (treated with test substance) | | | |
| 6.5 | 6.9 | 6.6* | 6.7* |
| 6.5 | 6.9* | 7.0 | 6.5* |
| 7.0 | 6.8 | 6.8* | 6.7* |
| 6.3* | 6.4* | 6.7* | 6.8 |
| 6.9* | 6.5 | 6.7 | 6.8* |
| Batch 2 (treated with water) | | | |
| 6.1 | 5.9 | 5.8* | 6.2 |
| 6.0 | 5.7* | 6.0 | 5.9 |
| 6.1 | 6.3 | 5.9* | 6.5 |
| 6.3 | 5.9 | 5.7 | 5.9 |
| 5.9 | 5.8 | 6.1 | 6.4 |

(a) For each group, calculate the mean and the 95% confidence limits. You should show all your working for the calculation of the confidence limits.

(b) Frame suitable null hypotheses concerning the effects of the weedkiller on (i) the growth and (ii) the mortality of the seedling.

(c) Test these hypotheses using appropriate statistical procedures.

*Northern Ireland A-Level Specimen 1990*

4. The table below shows the number of worms surfacing in two plots after $1m^2$ random quadrats were soaked with a dilute solution of potassium permanganate. Plot 1 had been regularly cultivated and manured, but Plot 2 had received no treatment at all for over two years.

| **Plot 1 - manured** | |
|---|---|
| **Quadrat** | **Worms / number per $m^2$** |
| 1 | 5 |
| 2 | 9 |
| 3 | 11 |
| 4 | 9 |
| 5 | 10 |
| 6 | 10 |
| 7 | 5 |
| 8 | 8 |

| **Plot 2 - untreated** | |
|---|---|
| **Quadrat** | **Worms / number per $m^2$** |
| 1 | 4 |
| 2 | 3 |
| 3 | 6 |
| 4 | 7 |
| 5 | 5 |
| 6 | 3 |
| 7 | 3 |
| 8 | 5 |

(a) State a null hypothesis that could be tested using this data.

(b) Carry out a *t*-Test to enable you to comment on the significance of the difference between the means of the two data sets.

*(Adapted from WJEC 1985 Biology)*

**Note**: In an actual investigation you might have to question the use of a *t*-Test here as the data are not really normally distributed.

5. Red blood cells (erythrocytes) transport oxygen from the alveolar surface to the respiring tissues. A group of students expressed the view that people living at high altitude should have higher red blood cell counts than people living at sea level.

The students selected two independent samples of people. Sample A contained nine people who lived at sea level and Sample B contained nine people who lived up a mountain at an altitude of 2000 m above sea level. Samples of blood were taken from each person and the cell counts determined using counting chambers.

(a) The table below shows the red blood cell counts of the nine people in each of the two samples.

| | **Number of red blood cells / $dm^3 \times 10^{12}$** | |
|---|---|---|
| Sample | A (Sea level) | B (High altitude) |
| 1 | 5.0 | 4.9 |
| 2 | 5.1 | 5.3 |
| 3 | 4.9 | 5.7 |
| 4 | 5.3 | 5.5 |
| 5 | 5.4 | 5.6 |
| 6 | 5.0 | 5.4 |
| 7 | 4.8 | 5.3 |
| 8 | 5.1 | 5.6 |
| 9 | 5.5 | 5.1 |

The data were analysed using a Mann-Whitney *U* test to test the null hypothesis that there is no difference in the red blood cell counts of the two populations at a 5% signficance level.

(i) The median value at high altitude (sample B) is 5.4. Find the median value for the population at sea level (sample A), and comment on the difference between the two median values. (2 marks)

(ii) Arrange the data from the table in order, in a form suitable for analysis using a Mann-Whitney *U* test. (2 marks)

(b) (i) For this investigation the critical value of U at the 5% significance level is 17. The values calculated for U are 18.5 and 62.5. Which value of U would you take to determine the significance of these results? (1 mark)

(ii) Do the results enable you to accept or reject the null hypothesis? Explain your answer. (2 marks)

(c) If the study was extended to use larger samples (100 people), explain how you would select the people for each sample. (3 marks)

*(London A and AS Paper 1992)*

6. Lichens are plant-like organisms which may grow on rocks, tree trunks and branches. After a preliminary survey of lichens in a small wood some students formulated the hypothesis that there was more growth of lichens on birch trees than on oak trees.

The students selected a sample of nine birch trees and nine oak trees and determined the percentage cover of three types of lichen (fruticose, foliose and crustose) in a single quadrat on each tree.

In order to standardise the collection of data, a quadrat frame (50 × 50 cm) was placed 1.5 m from the ground on the south facing side of each tree trunk. Only upright trees with a minimum diameter of 50 cm were selected.

(a) Choose three of the procedures followed by the students in order to standardise the collection of data, and explain why each procedure was followed. (3 marks)

(b) The table below shows the percentage cover of the three types of lichen as recorded in each quadrat on the nine trees.

| | **Percentage cover** | | | | | |
|---|---|---|---|---|---|---|
| | **Fructicose** | | **Foliose** | | **Crustose** | |
| **Tree** | **Oak** | **Birch** | **Oak** | **Birch** | **Oak** | **Birch** |
| 1 | 0 | 1 | 8 | 33 | 5 | 32 |
| 2 | 10 | 12 | 20 | 25 | 80 | 90 |
| 3 | 1 | 15 | 64 | 35 | 20 | 46 |
| 4 | 0 | 17 | 3 | 8 | 83 | 64 |
| 5 | 5 | 20 | 30 | 25 | 5 | 15 |
| 6 | 0 | 5 | 15 | 65 | 30 | 25 |
| 7 | 0 | 5 | 30 | 80 | 15 | 15 |
| 8 | 0 | 8 | 14 | 30 | 85 | 60 |
| 9 | 0 | 4 | 20 | 35 | 60 | 15 |

Data courtesy of Field Studies Council

The data were then analysed using a Mann-Whitney $U$ test to establish whether there was a difference between the percentage cover of each lichen on birch and on oak at a 5% significance level. The table below gives a summary of the results.

| **Lichen type** | **Tree** | **Median % cover** | **U value** |
|---|---|---|---|
| Fruticose | Oak<br>Birch | 0<br>8 | 72.5<br>8.5 |
| Foliose | Oak<br>Birch | $y$<br>33 | 62.5<br>18.5 |
| Crustose | Oak<br>Birch | 30<br>32 | 42.0<br>39.0 |

(i) Determine the median value for percentage cover of foliose lichens on oak trees. This is given as $y$ in the table. (3 marks)

(ii) Arrange the original data for percentage cover of the crustose lichens in a form suitable for analysis using a Mann-Whitney $U$ test. (2 marks)

(iii) For this investigation, the critical value of $U$ at the 5% significance level is 17. Which three values of $U$ would you take from the table to determine the significance of these results? (1 mark)

(iv) Do the results enable you to accept or reject the original hypothesis? Explain your answer. (3 marks)

*(London A-Level Specimen 1989)*

7. The table and the graph summarise the results of an investigation which measured the concentration of blood-lead in 15 pregnant women, none of whom showed any clinical symptoms of lead toxicity. In addition, the lead concentration of the umbilical-cord blood of each of these subjects was measured.

| **Blood-lead concentration/μmol dm$^{-3}$** | |
|---|---|
| **Fetus** | **Mother** |
| 0.65 ± 0.12 | 0.36 ± 0.07 |

Footnote 1. Figures above are means ± standard deviation
Footnote 2. p<0.01

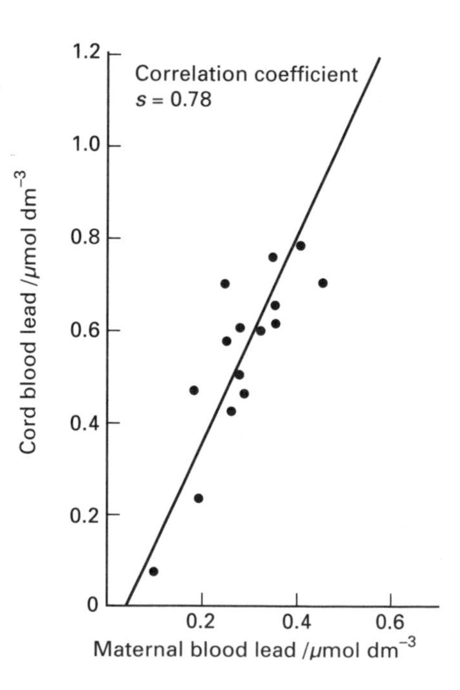

(a) Explain the meaning of the information given in the footnotes 1 and 2 under the table.

(b) (i) Referring to the graph, what is the relation between the maternal blood-lead level and the umbilical cord blood-lead levels?

(ii) Explain the meaning of the statistical information, correlation coefficient $r_s = 0.78$.

(c) (i) Explain what the data indicate concerning the passage of lead across the placenta.

(ii) Suggest a hypothesis to account for the higher concentration of lead in the umbilical-cord blood compared to the maternal blood.

(iii) Why should the safety level for blood-lead concentration in pregnant women be considered separately from that for non-pregnant women?

*(JMB AS-Level Human and Social Biology 1989)*

8. Assume that tongue-rolling in humans depends upon the presence of the dominant allele T, and the ability to taste phenylthiourea relies on the dominant allele P.
In a survey of 150 families in which both parents are heterozygous for both pairs of alleles, the numbers of their children of each phenotype are given below:

| | |
|---|---|
| Tongue-roller and taster | 240 |
| Tongue-roller and non-taster | 70 |
| Non-tongue-roller and taster | 64 |
| Non-tongue-roller and non-taster | 26 |

(a) (i) If the two genes show independent assortment, what would be the expected ratio of the four phenotypes represented by the children? (1 mark)

(ii) Using this ratio and the information in the question, what would be the expected numbers of the four phenotypes? (1 mark)

(b) Using the information given perform a chi-squared test ($\chi^2$) to determine whether there is a significant difference, at the 5% level of probability, between the observed and expected numbers of offspring. (6 marks)

Chi-squared formula:

$$\chi^2 = \sum \frac{(\text{Observed} - \text{Expected})^2}{\text{Expected}}$$

| Degrees of freedom | Probability (p) | | | | |
|---|---|---|---|---|---|
| | 0.50 | 0.30 | 0.20 | 0.05 | 0.01 |
| 1 | 0.455 | 1.074 | 1.642 | 3.841 | 6.635 |
| 2 | 1.386 | 2.408 | 3.219 | 5.991 | 9.210 |
| 3 | 2.366 | 3.665 | 4.642 | 7.815 | 11.241 |

*(AEB Human Biology Paper 1 1989)*

9. It has been suggested that, in humans, certain eye colour and hair colour are often inherited together. The table below shows the observed numbers (O) and the expected numbers (E) for each combination of eye and hair colour found in a sample of 130 British men.

| Eye colour | Hair colour: Fair | | Brown | | Black | | |
|---|---|---|---|---|---|---|---|
| | O | E | O | E | O | E | Totals |
| Blue | 65 | 53.3 | 26 | 32.0 | 8 | 13.7 | 99 |
| Brown | 5 | 16.7 | 16 | 10.0 | 10 | 4.3 | 31 |
| Totals | 70 | / | 42 | / | 18 | / | 130 |

The null hypothesis states that there is no link between the inheritance of eye colour and hair colour.

(a) (i) Test the null hypothesis by means of a $\chi^2$ test.

Fill in the values of $(O-E)$ and $(O-E)^2$ in the table below. (2 marks)

| Combination of eye and hair colour | $(O - E)$ | $(O - E)^2$ |
|---|---|---|
| Blue eyes with fair hair | | |
| Blue eyes with brown hair | | |
| Blue eyes with black hair | | |
| Brown eyes with fair hair | | |
| Brown eyes with brown hair | | |
| Brown eyes with black hair | | |

(ii) Use the formula given below to calculate the value of $\chi^2$. Show your working. (2 marks)

$$\chi^2 = \sum \frac{(O-E)^2}{E}$$

(iii) How many degrees of freedom are shown in this investigation? Explain how you arrived at your answer. (1 mark)

(b) For this number of degrees of freedom, $\chi^2$ values corresponding to important values of p are shown here.

| Value of p | 0.99 | 0.95 | 0.05 | 0.01 | 0.001 |
|---|---|---|---|---|---|
| Value of $\chi^2$ | 0.020 | 0.103 | 5.991 | 9.210 | 13.820 |

What conclusions can be drawn from this $\chi^2$ test concerning the inheritance of eye colour and hair colour? (3 marks)

*(London A-Level Paper 3 1992)*

10. In a certain population 4% of babies are born with a disease caused by a homozygous recessive gene. What proportion of the population are heterozygous carriers?

11. The frequency of one dominant allele is known to be 0.8. Only one other allele of this gene occurs. What percentage of the population would be homozygous for the recessive allele?

12. In humans, the ability to roll the tongue is dominant over the inability to do so. In a population where the frequency of the dominant allele is 0.7, the frequency of the heterozygous genotype is:
A 0.3: B O.21: C 0.42 or D 0.49. Which?

13. Some people (tasters) are able to taste a chemical called phenylthiourea while others (non-tasters) cannot. This ability to taste is determined by an allele T which is dominant to t. An analysis of a sample of people showed that 9% were non-tasters.

(a) Using the Hardy-Weinberg equations, calculate for this population
(i) the frequency of the t allele;
(ii) the frequency of the T allele;
(iii) the proportion of this population that is heterozygous for this gene. (4 marks)

(c) Before the Hardy-Weinberg equations can be applied, several assumptions must be made. Give three of these assumptions. (3 marks)

*(AEB A paper 1 1989)*

14. It is necessary to use the Hardy-Weinberg equation in this question. In this equation:

$$p^2 + 2pq + q^2 = 1$$

(where 1 is the whole population or gene pool)

$p^2$ represents the frequency of the homozygous dominant individuals
$2pq$ represents the frequency of the heterozygous individuals
$q^2$ represents the frequency of the homozygous recessive individuals
$p$ represents the frequency of the dominant gene
$q$ represents the frequency of the recessive gene
$p + q = 1$ (where 1 is the whole population or gene pool).

Phenylketonuria is a condition which leads to mental deficiency. It is controlled by one pair of alleles and is evident in the phenotype of a person who is a homozygous recessive.

Answer the following questions, using the equations given where necessary.

(a) In a population of 10 000 individuals, 25 are born with phenylketonuria.
(i) What is the frequency of the homozygous recessive genotype? (2 marks)
(ii) What is the frequency of the recessive gene? (2 marks)

(b) In medically advanced sectors of the world, clinical testing of new-born babies ensures that any homozygous recessives are detected and treated so that the genetic affliction is neutralised and these individuals will breed normally. Assuming random breeding with no other interfering factor, how many children of the next generation, consisting of 16 000 individuals would be phenylketonuric? (2 marks)

(c) By contrast, consider a population of 10 000 individuals where clinical testing is not available, so that the 25 homozygous individuals are not able to breed.
(i) How many recessive genes does this eliminate from the gene pool? (2 marks)
(ii) How many recessive genes are left in the gene pool? (2 marks)
(iii) What is the frequency of the recessive gene now that 25 individuals are excluded? (2 marks)
(iv) How many children will be phenylkentonuric if the following generation consists of 15 876 individuals? (2 marks)

(d) Comment on the significance of the number of phenylkentonuric children in the two contrasting populations. (2 marks)

(e) Suggest factors which will occur in human populations and which will interfere with statistics forecast by the Hardy-Weinberg equations. (2 marks)

15. (a) Give two assumptions that must be made when using the mark, release and recapture method to estimate population size. (2 marks)

(b) In a survey of a deer population, 80 deer were marked and released. Two weeks later a second sample was captured. Of these deer, 17 were seen to be marked and 3 were unmarked. Calculate the estimated population size. (2 marks)

*(AEB A Level 1993)*